STEM CELLS

New Frontiers in Science & Ethics

STEM CELLS

New Frontiers in Science & Ethics

editors

Muireann Quigley • Sarah Chan • John Harris

University of Manchester, UK

NEW JERSEY • LONDON • SINGAPORE • BEIJING • SHANGHAI • HONG KONG • TAIPEI • CHENNAI

Published by

World Scientific Publishing Co. Pte. Ltd.

5 Toh Tuck Link, Singapore 596224

USA office: 27 Warren Street, Suite 401-402, Hackensack, NJ 07601

UK office: 57 Shelton Street, Covent Garden, London WC2H 9HE

British Library Cataloguing-in-Publication Data
A catalogue record for this book is available from the British Library.

STEM CELLS
New Frontiers in Science & Ethics

ISBN-13 978-981-4374-24-8
ISBN-10 981-4374-24-5

Typeset by Stallion Press
Email: enquiries@stallionpress.com

Printed in Singapore by Mainland Press Pte Ltd.

Contents

Abbreviations

AD	Advanced Directive
AG	Advocate General
DNA	Deoxyribonucleic Acid
EU	European Union
FDA	Food and Drug Administration
(h)ESC(s)	(Human) Embryonic Stem Cell(s)
HFEA	Human Fertilisation and Embryology Authority
ICSI	Intra-cytoplasmic Sperm Injection
iPSC(s)	Induced Pluripotent Stem Cell(s)
IVF	*In Vitro* Fertilisation
IVG	*In Vitro* Gametogenesis
MP(s)	Member(s) of Parliament
NBAC	National Bioethics Advisory Commission
PGD	Pre-implantation Genetic Diagnosis
(i)SCNT	(Interspecies) Somatic Cell Nuclear Transfer
SC	Stem Cell
SCI	Spinal Cord Injury
UK	United Kingdom
US	United States

Acknowledgements

This edited collection owes much to the Wellcome Trust Strategic Programme Award which is held jointly by the Centre for Social Ethics and Policy and the Institute for Science, Ethics, and Innovation at the University of Manchester and the Centre for Research on Families and Relationships at the University of Edinburgh. The strategic award, The Human Body: Its Scope, Limits and Future, has supported the work of the editors of this collection and enabled them to carry out research on the ethics of stem cell science and other related areas.

We would like to thank all of our contributors, without whom we would not have such a great collection. In addition, we extend our thanks to our colleagues for their continued intellectual stimulation and to our families for their continued encouragement. We would like to offer our special thanks to one of our graduate research assistants, Nishat Hyder, whose help with formatting and the bibliography made our job much easier.

We would also like to acknowledge the assistance we have received from World Scientific Publishing in the preparation and completion of this edited collection, in particular Ms Sook Cheng Lim, whose patience and guidance has been much appreciated.

Acknowledgments

List of Contributors

Hannah Bourne
Oxford Uehiro Centre for Practical Ethics
University of Oxford
Suite 8, Littlegate House
St Ebbes Street
Oxford OX1 1PT

Sarah Chan
Centre for Social Ethics & Policy
Institute for Science, Ethics, & Innovation
School of Law
University of Manchester
Oxford Rd., Manchester, M13 9PL
United Kingdom

Sorcha Uí Chonnachtaigh
Centre for Professional Ethics at Keele (PEAK)
School of Law
Keele University
Staffordshire, ST5 5BG
United Kingdom
Email: s.ui.chonnachtaigh@keele.ac.uk

Sarah Devaney
Centre for Social Ethics & Policy
School of Law
University of Manchester
Oxford Rd., Manchester, M13 9PL
United Kingdom
Email: sarah.devaney@manchester.ac.uk

Katrien Devolder
Department of Philosophy and Moral Sciences
Bioethics Institute Ghent University
Blandijnberg 2, B-9000 Ghent
Belgium
Email: Katrien.Devolder@UGent.be

Thomas Douglas
Oxford Uehiro Centre for Practical Ethics
University of Oxford
Suite 8, Littlegate House
St Ebbes Street
Oxford OX1 1PT
Email: thomas.douglas@philosophy.ox.ac.uk

Catherine Harding
Oxford Uehiro Centre for Practical Ethics
University of Oxford
Suite 8, Littlegate House
St Ebbes Street
Oxford OX1 1PT

John Harris
Centre for Social Ethics & Policy
Institute for Science, Ethics, & Innovation
School of Law
University of Manchester
Oxford Rd., Manchester, M13 9PL
United Kingdom
Email: john.harris@manchester.ac.uk

Søren Holm
Centre for Social Ethics and Policy
Institute of Science, Ethics and Innovation
School of Law
University of Manchester
Oxford Rd., Manchester, M13 9PL
United Kingdom

Sheelagh McGuinness
Birmingham Law School
University of Birmingham
Edgbaston
Birmingham
B15 2TT
United Kingdom

Anna Pacholczyk
Institute for Science, Ethics, & Innovation
School of Law
University of Manchester
Oxford Rd., Manchester, M13 9PL
United Kingdom
Email: john.harris@manchester.ac.uk

Muireann Quigley
Centre for Social Ethics & Policy
Institute for Science, Ethics, & Innovation
School of Law
University of Manchester
Oxford Rd., Manchester, M13 9PL
United Kingdom
Email: muireann.quigley@manchester.ac.uk

Julian Savulescu
Oxford Uehiro Centre for Practical Ethics
University of Oxford
Suite 8, Littlegate House
St Ebbes Street
Oxford OX1 1PT

Loane Skene
Melbourne Law School
University of Melbourne
Victoria 3010
Australia
Email: l.skene@unimelb.edu.au

1

New Frontiers in Stem Cell Science & Ethics: Current Technology & Future Challenges

Muireann Quigley, Sarah Chan, and John Harris

This chapter introduces, and sets the context for, this edited volume. Some of the recent developments and debates in stem cell science are briefly reviewed. Then an overview of the chapters contained in *Stem Cells: New Frontiers in Science and Ethics* is given. Following this, the chapter moves on to indicate some future directions for debates regarding the ethics of stem cell science, highlighting a couple of areas which are likely to continue to generate ethical controversy in the coming years.

1. DEBATING STEM CELL TECHNOLOGIES

It is now over a decade since the first human embryonic stem cell (hESC) lines were produced.[1] In the years since Thomson's landmark achievement, the field of research on stem cells, both adult and embryonic, as well as cloning, gamete development and embryo manipulation, has expanded rapidly in many directions. Researchers have developed new techniques for manipulating stem cells; advances in basic research have

Centre for Social Ethics & Policy, Institute for Science, Ethics, & Innovation, School of Law, University of Manchester, Oxford Rd., Manchester, M13 9PL, United Kingdom. Email: muireann.quigley@manchester.ac.uk contributed to our understanding of stem cell biology, cell reprogramming and differentiation; clinical applications of stem cell therapies are under development in many areas. Meanwhile, the biotechnology industry has not been slow to recognise the potential of stem cells to generate marketable products, opening up questions over both the commercialisation of human biological materials and the proprietisation of science more generally. In this introductory chapter, we set this edited collection in context by briefly reviewing some of recent developments and debates in stem cell science. We then offer an overview of the chapters contained herein, before moving on to indicate a couple of areas which we believe will continue to generate controversy in the field in the coming years.

In terms of social and ethical issues, the use of human embryos in stem cell research has continued to generate as much controversy as ever. In addition, however, the new technological horizons opened up by more recent developments in stem cell science have led to novel ethical questions. Our understanding of the nature of the human embryo and its moral importance, of the process of reproduction, and of human nature and human identity, are being challenged by scientific advances that blur previously defined boundaries and undermine the clear distinctions that were once apparent. For example, as we will see below, the development of induced pluripotent stem cells (iPSCs) challenges the position of embryos or gametes as the only beginnings of new human life. Other advances, such as cytoplasmic hybrid embryos, blur the boundary between human and non-human.

Undoubtedly one of the most exciting developments in the stem cell field, in terms of both basic science and potential therapeutic applications, has been the discovery of methods for reprogramming somatic cells to a pluripotent, ESC-like state through the use of defined factors to produce iPSCs. Such reprogramming had long been conceived of as possible in theory, but dauntingly complex in practice. Nuclear reprogramming through somatic cell nuclear transfer (SCNT) (as used in the cloning of Dolly[2]) and through fusion with ESCs[3,4] was known to be possible, but it was not until 2006 that scientists reported the successful transformation of mouse fibroblasts into cells with pluripotent characteristics by the use of four gene factors.[5,6] These iPSCs, while not exactly equivalent to ESCs in their gene expression profile, demonstrated similar pluripotent properties and were able to contribute to the three primary embryonic germ layers (cell types that give rise eventually to all the different cell lineages of the body). This proof of principle in mouse cells was swiftly followed by the production of human iPSCs[7,8] a further step towards the eventual future use of this technique to generate patient-specific pluripotent cells for stem cell therapies.

The development of iPSCs was hailed as a milestone in stem cell biology,[9] opening up the possibility of producing pluripotent cells, capable of generating all sorts of differentiated cell types and tissues, from almost any cell of the body. It was also seen as a potential ethical breakthrough, since these pluripotent cells would be derived without the need to use human embryos or oocytes, providing an alternative to hESCs produced from regular embryos or SCNT. As we have pointed out, however, iPSCs raise a further set of ethical issues and questions over the moral status of the embryo.[10] For example, it has been shown that iPSCs are capable of contributing to tissues in the developing blastocyst and forming viable mouse embryos.[a,11] If all sorts of cells could now be used to produce embryos, what might it mean for the supposedly privileged status of the human embryo as the beginning of life if and when similar developments occur using human iPSCs? It is not yet the case that iPSCs will replace research on hESCs entirely, or at least not yet; important scientific questions remain to be answered that will require the ongoing use of hESCs.

Another important stem cell technology, and one that also leads to ethical issues of its own, is the use of oocytes from non-human animals in SCNT with human nuclei to create so-called cytoplasmic hybrid embryos.[12,13] Interspecies SCNT (iSCNT) between different non-human animal species has been used with varying degrees of success and has resulted, in some cases, in the production of viable embryos[14]; it has been suggested that iSCNT might find potential applications in conservation

[a] This requires iPSCs to be injected into blastocyst, since the iPSCs themselves cannot produce the outer layer of cells which form the placenta — see note 11.

biology as a way of cloning and preserving endangered species. This technique, however, is also used to produce hESCs using (most commonly) rabbit or bovine oocytes. This has attracted considerable scientific interest. A perceived advantage of this method is that it does not require the use of human oocytes, the procurement of which can raise ethical and practical difficulties. Yet the idea of creating embryos that are part-animal and part-human has provoked intense public reaction, tapping into deeply-held intuitions about the apparent importance of the species boundary between human and non-human animals and the ethical 'dangers' of transgressing this. Nevertheless, research on cytoplasmic hybrid embryos appears to hold some scientific utility, and it is likely that scientists will continue to pursue it; the ethical debate will, no doubt, also continue apace.

It is possible that the need for donated human oocytes in stem cell research might be addressed by another aspect of technological development. Much work in recent years has gone into exploring the possibility of *in vitro* gametogenesis, the production of sperm and oocytes from cells in culture. Both male and female gametes have successfully been produced from mouse embryonic stem cells,[15,16] leading to the possibility that this method might eventually be used to produce viable human gametes. This would clearly have important implications for reproductive medicine and the treatment of infertility in various situations, though raising further questions about how and whether the technology ought to be used in this context.[17,18] Together with the use of iPSC technology, it would in theory be possible to start with a somatic cell, such as a skin cell, and use this to produce pluripotent cells which could be differentiated to form gametes for reproductive use. Another application, however, might be in the production of potentially unlimited numbers of human oocytes for SCNT and hESC research, avoiding some (though not all) of the ethical issues associated with obtaining donated oocytes for research.

No discussion of developments in stem cell research over the past decade would be complete without mentioning the case of Woo Suk Hwang, the South Korean researcher who, in 2004, claimed to have successfully produced, for the first time, human embryonic stem cells from a nuclear transfer-derived blastocyst, so-called therapeutic cloning.[19] He followed this with a second publication a year later reinforcing this claim and reporting a ten-fold increase in the efficiency of his cloning method[20] — an apparent breakthrough that was hailed worldwide as a huge step towards therapeutic application of the technology and the creation of patient-specific stem cell lines. Alas for science (and *Science*), and for Hwang, it soon became apparent that his claims were false. The committee appointed by Seoul National University to review Hwang's work, following allegations first of unethical practice in obtaining egg donations and then of fraudulent manipulation of results, found that Hwang's data had "all been fabricated... [T]he research team of Professor Hwang does not possess patient-specific stem cell lines or any scientific bases for claiming having created one".[21]

Science immediately published an editorial retraction of the two papers.[22] In the wake of this scandal, Hwang was dismissed from his post at Seoul National University, and his licence to work on human embryos was revoked. He was indicted on charges of fraud, embezzlement, and breach of the South Korean Bioethics and Biosafety Act (2005). He faced criminal charges and possible imprisonment under the Act for a specific offence related to embryo research, but received a suspended sentence. It is probably the allegations and finding of fraud that have had the greatest impact on public perceptions of the case, and stem cell science in general. Thus, although Hwang's work, in this area at least, ultimately proved not to represent a real scientific advance in cloning and stem cell research, it remains a case study with important lessons about the conduct of science and the difficulties of ensuring research integrity, especially in high-profile areas such as stem cell science.

Since the Hwang case, the issue of human cloning, therapeutic or reproductive, seems to have faded somewhat into the background — perhaps eclipsed by the development of other technologies such as iPSCs — though work on SCNT continues. The derivation of ESCs from SCNT in non-human primates has been reported[23]; a further step towards replication of this work in humans. Successful reproductive cloning by SCNT, in non-human primates or in humans, has yet to be accomplished. On the former front, cloned monkeys have been produced from nuclear transfer using embryonic cells, and research into *in vivo* embryo development after SCNT is ongoing.[24] Regarding the latter, claims of attempted or actual human cloning occasionally crop up in the media, usually unverified and sometimes outlandish (the claims made by Severino Antinori and the Raelian sect being two notable examples). The idea of human reproductive cloning has of course attracted controversy and public attention — some might say disproportionately so, given how little scientific energy is being devoted to this in comparison to other areas of stem cell research; nevertheless, it remains a popular subject for debate, warranting ethical discussion.[25] However, the appearance in the published scientific literature of a genuine report of attempted human reproductive cloning[26] seems to have received surprisingly little attention, from ethicists or the public. It might be hoped that if, as seems at least possible, this research does one day result in successful human reproductive cloning, it would pass with similar lack of comment, being accepted and seen as uncontroversial, but this is unlikely to be the case.

The science of stem cells is thus in a phase of continued and fast-paced development, providing much excitement about new discoveries and the possibilities for science and medicine, as well as food for ethical thought.

Globally, legislation and policy regarding stem cells has also seen a period of considerable change, sometimes as a hasty attempt to respond to new scientific developments and the challenges they present to social attitudes, sometimes as the culmination of more gradual evolution of regulatory approach and scientific practice. In the United Kingdom, for example, when the birth of Dolly prompted a court case regarding the legality of human cloning,[27] the brief Human Reproductive Cloning Act 2001 provided a rapid legislative response. The 2008 revisions to the Human Fertilisation and Embryology Act saw a much more detailed consideration, after extensive consultation and debate, of a number of issues that had arisen in the years since the passage of the original Act in 1990. The new legislation included provisions regarding the use of embryo selection techniques for reproductive purposes, as well as for the use of embryos created by any method other than fertilisation, thus dealing with the cloning issue at greater length, and 'human admixed embryos', covering cytoplasmic hybrid embryos as well as the possibility of human–animal chimeric embryos. The legislation relating to embryonic stem cell research and cloning in Australia has undergone two rounds of change in the last decade, with the establishment of a consistent national legislative framework through the Research Involving Human Embryos Act 2002 and Prohibition of Human Cloning Act 2002, enacted at federal level and incorporated into state legislation. The legislation was reviewed, as required by its own provisions, in 2005[b]; another review, similarly required, was completed in 2011.[28]

2. STEM CELLS: SOME ETHICAL AND LEGAL ISSUES

In this introductory chapter, we have already seen that stem cell science moves quickly. The social and legal landscape in which it operates is ever changing, sometimes struggling to keep pace with the evolution of an area which pushes scientific frontiers and potentially threatens to cross moral boundaries. The chapters in this edited collection each deal with difficult aspects of this often controversial science. We could not hope to cover all areas of interest and dispute in relation to the use of stem cells; to do so would require several volumes. Even so, a diverse range of issues is examined in the chapters making up this volume. They explore ethical, legal, and social concerns and arguments regarding the use of stem cells in a variety of contexts, including for research, the treatment of disease, and in relation to infertility.

One of the areas of stem cell science which continues to fuel ethical debate is the use of embryonic stem cells for research and in clinical applications. For this reason, the two chapters which comprise Section 2 of the book aim to situate the embryo within the stem cell debate. In the first of these, Sorcha Uí Chonnachtaigh discusses the monopoly that moral status

---

[b] This review, undertaken by the Lockhart Committee, resulted in the passage of the Prohibition of Human Cloning for Reproduction and the Regulation of Human Embryo Research Amendment Act 2006.

seems to have in debates on human embryonic stem cell (hESC) research. She argues that focusing on the concept of moral status results in our field of moral concern being unduly and mistakenly narrowed. In order to explore why this is the case, she examines some of the main approaches and positions on moral status. She notes that each of these is problematic in different ways and outlines the reasons for this. She concludes that threshold concepts of moral status are to be preferred over non-threshold ones. However, when drafting policy regarding hESC research, there may be other reasons to give embryos serious consideration, since the point of public policy is not to enforce or privilege any particular individual's or group's private morality.

Following on from this, in Chapter 3, Sheelagh McGuinness argues that the 'embryo' is a constructed concept; it has varied meanings which are at times manipulated in different contexts. In a detailed analysis of these contexts, she demonstrates how the embryo is a legally, morally, socially, politically, and culturally constructed entity. Recognition of this is important if we are to engage in an informed and nuanced debate about the use of embryonic stem cells and hESC research. The reason for this, McGuinness maintains, is that these constructions influence what we come to deem as being (im)permissible with regards to the use of the embryo. This in turn can have a powerful effect on how we regulate embryology and different types of embryos, the 'research embryo', the 'reproductive embryo', the 'hybrid embryo', etc. Attaining an understanding of these different constructions can, she argues, help us to understand disagreements between parties and how to take account of this in policies, such as those affecting the use of stem cells.

The chapters comprising the next section of this edited collection examine some of the legal issues regarding stem cell research and technology. The section starts with a chapter by Loane Skene which looks at the regulation of human stem cell technologies. She begins by outlining some of the recent developments and advances which have taken place in stem cell science and treatment and then moves on to explore the regulatory landscape in light of this scientific progress. In particular, she notes the regulatory implications of the use of induced pluripotent stem cells (iPSCs) in research and the impact that this might have on the use of hESCs. She argues that the use of iPSCs is not without ethical concerns. In the final part of the chapter, Skene looks at policy issues regarding the use of human oocytes in research, the creation of human–animal hybrids, and so-called 'three parent' embryos.

Sarah Devaney continues the legal focus in the next chapter, where she asks whether human embryos created through *in vitro* fertilisation (IVF), but intended for use in stem cell research, could be usefully dealt with by a legal property framework. In making her analysis, she examines the current approach of English law to the progenitors of IVF embryos. She argues that, in practical terms, legislative provisions have the effect of allowing progenitors to shape the legal status of embryos and, in doing so, bestows upon them property-type powers over those embryos. Devaney then presents reasons why, once they have been made available for stem cell research, embryos should be considered as property and their providers compensated accordingly. Before bringing the chapter to a close, she also examines the challenges for the regulatory sphere in trying to protect the interests of those who provide embryos for stem cell research.

The fourth part of this book consists of three chapters in which the analysis moves on to dissect out some ethical distinctions and inconsistencies found in the debate on stem cells. In the first of these, in Chapter 6, Katrien Devolder argues against the created–discarded distinction in hESC research. She notes that those who support this supposed distinction draw a moral line between those embryos left over from IVF treatment and those created explicitly for research purposes. The moral line drawn places the different embryos on opposing sides in terms of the ethical permissibility of their creation: It is permissible to use those created for IVF in stem cell research, but not to create embryos solely for this purpose. Devolder argues against the created–discarded distinction by demonstrating how the arguments justifying the destruction of IVF embryos used by proponents of the distinction also apply to embryos created for research. Following this, she explores whether the distinction can be rescued. She looks at, among others, the 'nothing-is-lost' argument, which is often deployed in defence of the distinction. But she concludes, nonetheless, that in respect of the embryos used in stem cell research, these arguments cannot do the work expected of them in supporting a discarded–created distinction.

In Chapter 7, Muireann Quigley picks up on one of the themes touched on in Devolder's chapter and discusses the notion of complicity in stem cell research. In particular, she asks whether countries which ban hESC research should benefit from hESC-derived therapies if and when they become available. She argues that while these countries do not commit the purported principal wrong of embryo destruction themselves, they do act with the effect that embryo destruction occurs. When they import hESC-derived therapies, they act in a manner that creates an incentive for the moral wrong to continue in the future; they encourage the enterprise of stem cell research as a whole. Quigley admits that complicity is a complex notion and that, in the context of hESC research, it runs into the inherent problems of collective action and responsibility. Despite this, she concludes that for their own internal moral regulation, countries that object to hESC research on the grounds of embryo destruction should refrain from importing hESC-derived therapies once available.

Chapter 8 from Søren Holm is the final chapter in this part of the book. It focuses on claims of hypocrisy in stem cell research. Holm observes that certain commentators claim there is hypocrisy in the position of those who oppose stem cell research. To begin, he examines whether inconsistency and hypocrisy are intractably linked. He argues that being hypocritical is more than mere inconsistency and that we cannot impute hypocrisy from this alone; hence while people might be inconsistent within the stem cell debate, true hypocrisy is rare. He goes on to analyse whether benefiting from any past wrongdoing incumbent upon stem cell research renders one a hypocrite. In this respect, he concludes that those who make such claims cannot readily fit the claims into their own ethical theory. Finally, Holm tackles the issue of dissenting persons within democratic states and argues that contentious issues within stem cell research ought to be resolved in ways compatible with living in a democracy.

The last and final part of this edited collection focuses on new developments in stem cell science and explores whether these developments require new frontiers in the ethical arguments relating to stem cells. In Chapter 9, Thomas Douglas, Catherine Harding, Hannah Bourne, and Julian Savulescu look at the potential for stem cell research to facilitate same sex reproduction. The authors suggest that in the future, it may be possible to create gametes from human stem cells, and that when this comes to pass, it could open up another avenue for treating infertility. They call this *in vitro* gametogenesis (IVG). Specifically, IVG could allow same-sex couples to have children genetically related to both partners. They examine four different arguments to see whether such technology could justifiably be denied to same-sex couples. The first of these looks at arguments which claim that same-sex couples have chosen to be infertile. Then they examine whether IVG might count as treatment for different-sex couples, but enhancement for same-sex couples. They follow this with an examination of the argument from unnaturalness and the claim that it is unnatural to create children with same-sex genetic parents. The final argument they explore focuses on whether children created through same-sex IVG could be expected to have a lower level of well-being than those born to different-sex parents. The authors conclude that none of the lines of argument pursued provide robust or justified reasons for denying this stem cell-based technology to same-sex couples once it is available and that to do so would be highly discriminatory.

In the tenth and last chapter in this volume, Anna Pacholczyk and John Harris move the discussion and analysis to examine clinical research involving stem cells. They are particularly interested in some of the ethical issues regarding participation in early stage clinical stem cell trials. Specifically, they look at the case of individuals with an acute spinal cord injury (SCI). They investigate whether and in what ways patients in the acute stage of injury might be considered to be vulnerable. They also ask whether this vulnerability is a strong enough reason to exclude them from early-phase stem cell trials. In this context, Pacholczyk and Harris explore the general problem of therapeutic misconception in relation to phase I clinical trials, expressly addressing concerns about the possibility that vulnerable SCI patients might believe that they are receiving a proven, albeit novel, medical treatment. They argue that even though we might have reasons to be concerned about the participation of acute-stage SCI patients in phase I clinical stem cell trials, there are not robust reasons to suppose that they are not capable of making the decision to take part.

3. FUTURE CHALLENGES AT THE NEW FRONTIERS OF SCIENCE?

Stem cell science and research continues to fascinate and disturb in equal measure. The future promises to be as interesting as the present. One important consequence of stem cell research broadly conceived, the full effect of which has yet to be felt, is the gradual, and probably inevitable, erosion of certain moral categories concerning living things. In particular, we see two areas of research as providing fertile ground for ethical debate in the coming years, both of which challenge our ideas regarding moral boundaries. The first of these will be the continued development of the science in relation to iPSCs. The second will be the future direction of stem cell research which utilises human–animal embryos, such as chimeric and cytoplasmic hybrid embryos.

The ethical challenges regarding the first of these can be seen clearly when we consider how future research might build on the advances already made regarding iPSCs. As a number of essays in this volume eloquently attest, we are already seeing the consequences of developments regarding the plasticity of these cells. The ability to induce cells to alter their type, literally to change what they can do, may have far-reaching consequences. As we mentioned earlier in this chapter, somatic cells which are specialised, for example skin cells, can now be reprogrammed to express certain genes which renders them pluripotent; that is, they can be used to generate any cell type. We also saw that viable mouse embryos have been created using this technology. If and when it becomes possible to create embryos from reprogrammed human somatic cells, it may no longer be possible to ethically differentiate between iPSCs and hESCs on the grounds that iPSCs do not have the capacity to become human embryos. Thus, in so far as the embryonic stage of development has certain normative resonance for, and concerning, particular creatures, namely humans, this resonance may become itinerant.[29] The reason for this is because arguments regarding the potential of the human embryo to become a person are used by many to confer special moral status on the embryo. This potentiality could be seen as being understood by some as affording protection and moral status to *whatever* has the potential to grow into a normal adult human being. As such, future scientific developments could have implications for the moral commitments of those who defend such a view, since every human cell, in having the potential to become an embryo, might be entitled to protection. For this potential, which is claimed to be important, would not only be possessed by embryos properly so-called, but by whatever has the potential to become an embryo or a zygote (or to return to being one).[30]

In making this claim, however, we must differentiate between 'potential' as meaning merely or technologically possible and 'potential' in terms of a normative moral account of potentiality. Of course, once it is technologically possible, all human cells will have the 'potential' to become a human embryo. Yet, this by itself does not tell us much about the normative arguments regarding this potential nor the moral commitments of those who assign moral importance to it. In order to do this, we would need a normative account of potentiality to work with. There is not the space here to go into much depth on this issue and its relevance for further iPSC research. However, we have two brief points that we wish to make.

The first, as Holm has noted, is that embryonic (embryo-like) cells and embryos are not the same sort of things.[31] Think of the technology in relation to iPSC mice: Only under a particular set of circumstances can a somatic (mouse) cell be induced to exhibit pluripotency, and only under a further set of circumstances, by injecting the iPSCs into an already formed blastocyst, can it become an embryo which can give rise to the live birth of a mouse. Thus, a very specific set of conditions must obtain in order for somatic cells to become (or in fact contribute to the formation of) an embryo. If the technology with regard to human somatic cells follows the mouse model, then these very particular conditions would also need to apply before we get a human embryo from these cells. Like the somatic mouse cells, it would be a two-stage process where embryonic-type cells are created first (iPSCs), followed by the formation of an actual embryo. Thus, making embryonic-type cells in the form of iPSCs is not the same as making embryos; and creating the former need not entail creating the latter.

However, for those who wish to defend a normative account of potentiality as determining the moral status of embryos, the fact that the iPSCs themselves are not embryos would not seem to get them around the issue of the moral status of *actual* iPSC-derived embryos. If one believes that human embryos have a special moral status by virtue of their potential to become a person, one might simply bite the iPSC bullet. Thus, although those who object to hESC research were among the first to welcome the development of iPSCs as an ethical source of stem cells, in the light of new scientific advances, they might want to protect the human embryo whatever its genesis. The point, related to Holm's above, is that a commitment to protecting the embryo does not always necessitate a commitment to protecting its precursors, be they gametes or somatic cells. And it is this which leads us on to our second comment regarding potentiality and both hESCs and iPSCs.

The argument just made does not mean that those who subscribe to the potentiality view are secure in the ethical claims contained therein; far from it. They would still need to give a normative account of the moral significance that the 'potential' to become a person carries. They would need to explain why the potential of the human embryo is different from the potential of its precursor gametes. A tempting response might be to claim that it is only the fused gametes under particular conditions which have the relevant potential; if we do not bring about the conditions necessary for the fusion of those gametes then no embryo will result. Yet when thinking about IVF embryos, it is also true that without creating the very specific circumstances in which the embryos get transferred to a uterus in order to undergo gestation, no person will ever result. Consequently, if a distinction is to be drawn regarding the potential of embryos as opposed to gametes, a robust argument is needed to explain why environmental and situational facts are applicable and morally relevant for the former, but not the latter. Likewise, for iPSC-derived embryos, a justification would be required if one is to defend protecting the embryo because of its potential, but not the precursor iPSCs in virtue of theirs. Whether we are talking about gamete and cell precursors or the embryos themselves, all require distinct sets of conditions for any potential to be realised.

The second strand of stem cell science which we expect will continue to generate much ethical controversy and debate is the use of human–animal interspecies (humanimal) embryos like chimeras and cytoplasmic hybrids. The prospect of mixing species exercises both popular and scientific imagination in different ways. As the science progresses, the ethics and permissibility of creating hybrids, chimeras, and other forms of novel combinations of cells may become one of the most fascinating problems facing contemporary science and bioethics. It has already become obvious that not only can we learn much regarding basic science from the creation of such ambiguous entities, but the therapeutic prospects in this area offer immense promise for the treatment of disease and for the amelioration of the human condition.[32]

Many have assumed that there exists a barrier between species which protects their integrity. The creation of humanimal embryos questions this, prompting us to enquire into the point and purpose of the species barrier, and to ask what reasons might justify breaching it. The existence of such embryos prompts us to question, in dramatic form, the use of the prefix 'human' in so many ethical and political contexts, inviting us to contemplate whether we can continue to talk of human rights, human dignity, humankind, and human values. It could be argued that what distinguishes human and animal cells is not so much their origin, their provenance one might say, but what they do and what sorts of creatures they end up being (part of).[33] If this is correct, then we would have reason to think that the so-called 'species barrier' between humans and other species may erode.

It is possible, as some fear, that current and future developments such as the ones just discussed might lead to a diminution in respect for human life. Yet we think it more likely that the science should prompt us to give increased attention to discussing, debating, and analysing exactly what it is about life and living creatures that commands moral respect in the first place. Fundamental questions about the value of life are what have made contemporary bioethics so interesting and important. They are also what make the field of stem cell research so vital, both literally and metaphorically. We hope that this vitality is evident in the contributions which follow in this volume.

REFERENCES

1. Thomson JA, Itskovitz-Eldor J, Shapiro SS, *et al.* (1998) Embryonic Stem Cell Lines Derived from Human Blastocysts. *Science* **282(5391):** 1145–1147.
2. Wilmut I, Schnieke AE, McWhir J, *et al.* (1997) Viable Offspring Derived from Fetal and Adult Mammalian Cells. *Nature* **385(6619):** 810–813.
3. Tada M, Takahama Y, Abe K, *et al.* (2001) Nuclear Reprogramming of Somatic Cells by *In Vitro* Hybridization with ES Cells. *Current Biology* **11(19):** 1553–1558.
4. Hochedlinger K, Jaenisch R. (2006) Nuclear Reprogramming and Pluripotency. *Nature* **441(7097):** 1061–1067.
5. Takahashi K, Yamanaka S. (2006) Induction of Pluripotent Stem Cells from Mouse Embryonic and Adult Fibroblast Cultures by Defined Factors. *Cell* **126(4):** 663–676.
6. Reviewed in Geoghegan E and Byrnes L. (2008) Mouse Induced Pluripotent Stem Cells. *International Journal of Developmental Biology* **52(8):** 1015–1022.
7. Takahashi K, *et al.* (2007) Induction of Pluripotent Stem Cells from Mouse Embryonic and Adult Fibroblast Cultures by Defined Factors. *Cell* **131(5):** 861–872.
8. Yu J, *et al.* (2007) Induced Pluripotent Stem Cell Lines Derived from Human Somatic Cells. *Science* **318(5858):** 1917–1920.
9. Vogel G, Holden C. (2007) Developmental Biology. Field Leaps Forward with New Stem Cell Advances. *Science* **318(5854):** 1224–1225.
10. Chan S, Harris J. (2008) Adam's Fibroblast? The (pluri)potential of iPCs. *Journal of Medical Ethics* **34(2):** 65–66.
11. Wernig M, *et al.* (2007) *In Vitro* Reprogramming of Fibroblasts into a Pluripotent ES Cell-like State. *Nature* **448(7151):** 318–324.
12. Academy of Medical Sciences. (2007) *Inter-species Embryos.* Available at http://www.acmedsci.ac.uk/p48prid51.html#downloads (Accessed 16 June 2011).

13. Skene L, Testa G, Hyun I, *et al.* (2009) Ethics Report on Interspecies Somatic Cell Nuclear Transfer Research. *Cell Stem Cell* **5(1):** 27–30.
14. Beyhan Z, *et al.* (2007) Interspecies Nuclear Transfer: Implications for Embryonic Stem Cell Biology. *Cell Stem Cell* **1(5):** 502–512.
15. Geijsen N, *et al.* (2004) Derivation of Embryonic Germ Cells and Male Gametes from Embryonic Stem Cells. *Nature* **427(6970):** 148–154.
16. Hubner K, *et al.* (2003) Derivation of Oocytes from Mouse Embryonic Stem Cells. *Science* **300(5623):** 1251–1256.
17. Testa G, Harris J. (2005) Ethics and Synthetic Gametes. *Bioethics* **19(2):** 146–166.
18. Mathews D, Donovan PJ, Harris J, *et al.* (2009) Pluripotent Stem Cell-derived Gametes: Truth and Potential Consequences. *Cell Stem Cell* **5(1):** 11–14.
19. Hwang WS, Ryu YJ, Park JH, *et al.* (2004) Evidence of a Pluripotent Human Embryonic Stem Cell Line Derived from a Cloned Blastocyst. *Science* **303(5664):** 1669–1674.
20. Hwang WS, Ryu YJ, Park JH, *et al.* (2005) Patient-specific Embryonic Stem Cells Derived from Human SCNT Blastocysts. *Science* **308(5729):** 1777–1783.
21. As quoted in Weissmann G. (2006) Science Fraud: From Patchwork Mouse to Patchwork Data. *The FASEB Journal* **20:** 587–590.
22. Kennedy D. (2006) Editorial Retraction. *Science* **311(5759):** 335.
23. Byrne JA, *et al.* (2007) Producing Primate Embryonic Stem Cells by Somatic Cell Nuclear Transfer. *Nature* **450(7169):** 497–502.
24. Sparman ML, Tachibana M, Mitalipov SM. (2010) Cloning of Non-human Primates: The Road "Less Travelled By". *International Journal of Developmental Biology* **54(11–12):** 1671–1678.
25. Harris J. (2004) *On Cloning.* Routledge, London.
26. Zavos PM, Illmensee K. (2006) Possible Therapy of Male Infertility by Reproductive Cloning: One Cloned Human 4-cell Embryo. *Archives of Andrology* **52(4):** 243–254.
27. *R (Quintavalle) v Secretary of State for Health* [2001] EWHC 918 (Admin).
28. See https://legislationreview.nhmrc.gov.au/2010-legislation-review (Accessed 16 June 2011).
29. See Harris J. (2007) *Enhancing Evolution.* Princeton University Press, Princeton and Oxford, Chapter 10.
30. See Marquis D. (1989) Why Abortion is Immoral. *Journal of Philosophy* **86(4):** 183–202; Savulescu J. (2002) Abortion, Embryo Destruction, and the Future of Value Argument. *Journal of Medical Ethics* **28**: 133–35; and Marquis D. (2005) Savulescu's Objections to the Future of Value Argument. *Journal of Medical Ethics* **31**: 119–122.

31. Holm S. (2008) Time to Reconsider Stem Cell Ethics: The Importance of Induced Pluripotent Cells. *Journal of Medical Ethics* **34**: 63–64, p. 63.
32. See generally Academy of Medical Sciences (2007) *Interspecies Embryos.* Available at http://www.acmedsci.ac.uk/p48prid51.html#downloads (Accessed 16 June 2011).
33. Harris J. (2011) Taking the 'Human' Out of Human Rights. *Cambridge Quarterly of Healthcare Ethics* **20(1):** 9–20.

2

The Monopoly of Moral Status in Debates on Embryonic Stem Cell Research

Sorcha Uí Chonnachtaigh

In academic and public debates about the moral acceptability of embryonic stem cell (ESC) research, there is a marked tendency to focus on the issue of the moral status of the human embryo. While problematically narrow, the almost universal dominance of the concept of moral status indicates that it warrants consideration. The importance of the human embryo's moral status in debate relies on an assumption that moral status is significant and possibly determinative with regard to how we may treat it. Moral status means different things to different people in both public/political debate and academic argument. I would dispute the utility of the concept of moral status unless it is to be applied as a threshold. Even then, focus on the issue of moral status mistakenly narrows the field of moral concern. In order to tease this out, a number of prominent approaches that represent the range of positions on moral status will be examined: the Conception View, the Personhood View and the Intermediate Views.

Centre for Professional Ethics at Keele (PEAK), School of Law, Keele University, Staffordshire, ST5 5BG, United Kingdom. Email: s.ui.chonnachtaigh@keele.ac.uk.

1. INTRODUCTION

In academic and public debates about the moral acceptability of human embryonic stem cell (hESC) research, there is a marked tendency to focus on the issue of the moral status of the human embryo.[a,1] That this is the case in countries with different historical and socio-cultural contexts is quite remarkable. While I would argue that this 'monopoly of moral status' is problematic because it neglects other morally important issues, its almost universal dominance indicates that moral status warrants consideration. The continued controversy surrounding the question of the embryo's moral status further supports an examination of the concept and may explain its dominance (though this is not something I will attempt to explain).

The question of the human embryo's moral status relies on an assumption that moral status is significant and possibly determinative with regard to how we may treat it. Of course, moral status means different things to different people in both public/political debate and academic argument. I would dispute the utility of the concept of moral status unless it is to be applied as a threshold. Even then, the concept mistakenly narrows the field of moral concern.[2] In order to tease this out, I will examine a number of prominent approaches that represent the range of positions on moral status: the Conception View, the Personhood View and the Intermediate Views.

2. THE CONCEPTION VIEW

The conception view, put simply, holds that the human embryo has full moral status from conception. There is no single conception view, but many. Many conception accounts of moral status have a theological nature

[a] I am, by no means, the first to notice the excessive focus on the moral status of the embryo. For example, Suzanne Holland (see note 1) argues for a feminist analysis of new technologies that would take account of the broader social context, in which the "existing patterns of oppression and domination in society" (Sherwin in Holland (see note 1, p. 73)) are taken into account. There are other issues relating to embryonic stem cell research that ought to be considered.

or religious origins, for example the theological position of the Roman Catholic Church. Most arguments for the conception view rely on some or all of the following concepts: genetic individuality, human dignity, and sanctity of life. (Some are also reinforced by reference to secondary principles such as the precautionary principle and the potentiality principle.) While the arguments in favour of assigning moral status to the embryo from conception are weak, the onus falls on those who do not agree with them to make the counter-argument, given their historical primacy.[3]

2.1 The Roman Catholic Position

It is sometimes presumed (by non-Catholics) that the Catholic position is absolutist; that the Church believes a human being with full moral status (or right to life or personhood, depending on the framework one uses) exists from fertilisation. The more theological way of framing this would be to contend that the rational soul is present in the human organism from conception. Such a position can be articulated in the latter, religious way, but is more often articulated in a secular way.[4] They essentially mean the same thing. However, the Catholic position is more nuanced than this even if, or perhaps because, it is ambiguous.

Donal Murray asserts that the Church's position is a nuanced one. Neither the Church nor Catholic theologians argue that the embryo has absolute moral status based on the presence of a soul from conception, though this is often the common perception.[b,5,6] Because of scientific ambiguity, the Church accepts that we cannot know that a soul is present from conception. However, the Church does argue that the embryo either has a soul/full moral status from conception or it *probably* has a soul (and full moral status) from conception:

> Even if the presence of a spiritual soul cannot be ascertained by empirical data, the results themselves of scientific research on the human

[b] See Eberl (note 6) for a more academic overview of the Thomistic account of moral status, on which much of the Catholic position is based.

> embryo provide "a valuable indication for discerning by the use of reason a personal presence at the moment of the first appearance of a human life: how could a human individual not be a human person?" … [W]hat is at stake is so important that, from the standpoint of moral obligation, the mere probability that a human person is involved would suffice to justify an absolutely clear prohibition of any intervention aimed at killing a human embryo.[7]

There are contentious assumptions made in Pope John Paul II's statement concerning individuality from conception and the equation of human life with human beings and persons. However, we can also see that whether the human zygote is ensouled or not is something the Church does not pretend to assert, instead the Church opts for a precautionary position based on the entity's undisputed humanity. This is problematic for many reasons. First, there is the main underlying presumption of the moral superiority of the human species.[c] Second, there is a reliance on the precautionary principle rather than on moral status itself, that is to say there is no appeal to relevant features of moral status holders (other than human species membership). This account ultimately relies on a special metaphysical belief, namely belief in the soul (an immaterial essence or quality) without rational support for its existence. The existence of the intangible soul cannot be proven. Lack of proof does not mean that the soul does *not* exist, just that it cannot be a necessary element in rational argument regarding moral status.

2.2 Human Dignity

In her now (in)famous editorial, 'Dignity is a Useless Concept', Ruth Macklin[8] argues that human dignity is too often used as a mere slogan, or in the place of other more precise terms.[9] Dignity does indeed have many meanings (both historical and contemporary[10,11]). The most philosophical

[c] This is now commonly referred to as 'speciesism' in the academic literature. The term came into common usage in the 1970s with writers such as Peter Singer, Tom Regan and Jonathan Glover defining it as a morally arbitrary distinction between species — most often between the human species and other animal species.

account of dignity used as a basis for arguments against ESC research is found in Immanuel Kant's (1785/1997) *Groundwork of the Metaphysics of Morals:*

> So act that you use humanity, whether in your own person or in the person of another, always at the same time as an end, never merely as a means.[12]

Patrick Kain claims that Kant believed moral status applied to all human beings, based on a careful analysis of Kant's biological and psychological theories.[13] However, others argue that careful interpretation of Kant's use of the terms 'humanity' and 'person' reveal that he was not using humanity to connote a biological category or membership of the species *Homo sapiens* but to refer to a capacity for reason, and that 'person' refers only to beings with this capacity.[14,15] Kant refers to respecting humanity, but in order to do so, we must respect it *in persons.* Given that Kant refers to a person as a "subject whose actions can be *imputed* to him",[16] it can be clearly deduced that Kant identifies a person as a being with certain capacities, and not with certain biological characteristics. This would imply that human embryos fall outside the category of the Kantian person.

Deryck Beyleveld and Roger Brownsword[17] offer a more contemporary account of human dignity as it relates to bioethics and law. Beyleveld and Brownsword work within a rights framework as can be seen from their adoption of a modified Gewirthian principle of generic consistency "locat[ing] the essence of dignity in agency".[18] However, Beyleveld and Brownsword diverge from Gewirth in rejecting the possibility of partial agency. They hold that one is either an agent or one is not an agent.[19] While they do allow for potential agency, they argue that an agent can only have duties *to* other agents, and a potential agent is necessarily a non-agent.[20] Duties are owed *in relation to* possible agents "in proportion to the degree to which they display evidentially the necessary capacities and characteristics of agents"[21] such that possible agents have a derivative, rather than intrinsic, moral status. Based on this assertion, it would be more appropriate to place Beyleveld and Brownsword in the Intermediate View of moral status category than in the Conception View.

This is reinforced by Beyleveld:

> Unless agents have a compelling reason to consider the embryo-fetus [sic] to be an agent from the moment of conception as they have to consider any adult human to be an agent, Gewirthians must reject the 'pro-life' position — which usually rests on the idea that human life (biologically defined) is the sufficient condition for having full moral status.[22]

2.3 Genetic Individuality/Uniqueness

The argument that moral status should be ascribed based on the genetic uniqueness (or individuality) of the human embryo, which is generated at conception, involves the implied claim that uniqueness and individuality are morally significant markers of moral status. Basing moral status on genetic individuality is highly problematic. Quite simply it fails to acknowledge the fact of monozygotic twins, who would be two separate individuals with moral status despite being genetically identical. To suggest that identical twins have diminished moral status because they are not genetically unique individuals would be absurd.

We must address the question of individuality as a basis for moral status separately. To assert that all entities with moral status are individuals, and therefore all individuals have moral status is an inadequate claim. Individuality could be conceived of as an arbitrary feature or, at most, a necessary but insufficient feature of an entity with moral status.

It is the final feature ('specialness') that seems to be the crucial point for those who use genetic uniqueness as a basis for ascribing moral value. The fertilisation of an egg with sperm creates a new zygote with the genesis for a human being that has never before existed and will never again exist. There is an assumption that human beings are morally special — for we are not concerned with non-human animals that are similarly genetically unique (in the chronological sense described above). The idea here is that every single possible human is precious and worth protecting. Of course, if my parents had sexual intercourse the day after I was conceived rather than the day I was conceived, under circumstances that were similarly conducive to reproduction (i.e. mother was ovulating and father was similarly fertile), they would very likely have conceived a different human being, and I would never have come into existence. If they had chosen not to have intercourse at the time they did, I would not exist, and I cannot think of anyone who would consider this a tragedy.[d] If we want to bring about every possible genetically unique human being, we would have to be extremely careful with all human gametes and make sure that as many healthy gametes are enabled to merge and create new zygotes, and provide suitable environments to bring the embryos to term. This is absurd; while we might value other human beings in part due to the diversity of personalities and manifestations of the human that they represent, this does not mean that the moral status we ascribe to them is based only on genetic uniqueness.

Arguments based on human dignity or genetic uniqueness are ultimately unconvincing as grounds for ascribing full moral status to the human embryo. However, they do point to a widespread and long-standing desire to protect human life. Whatever the flaws in the approaches outlined above, and despite rational arguments in favour of alternatives, it is not unreasonable to think that there is a value in the species that ordinarily gives rise to beings with full moral status.

3. THE PERSONHOOD VIEW

The quote below from Søren Holm (a critic of this approach) gives an arguably simplified and basic sketch of what he calls 'the standard liberal approach' to the moral status of the human embryo. This is, nevertheless, a useful starting point from which to view the personhood account that follows:

> By analyzing [human embryos'] characteristics we can see that they are not persons and that it is not wrong to kill them, and we can also see that

[d] I would like to think that my family and friends are glad to know me, but they would never have known me in the scenario sketched here.

> they do not qualify for any kind of 'lower' moral status (for instance, based on sentience). Given that a few other provisos are fulfilled (e.g., concerning permission from the owners of the embryo or gametes), it can be shown that derivations of stem cells is morally acceptable, perhaps even mandatory.[23]

This general description encapsulates both the central significance of personhood and the concept of personhood acting as a threshold (rather than a concept that admits of degrees) common to liberal–permissive views of moral status.[e] In this section, I will focus on the personhood account developed and championed by John Harris.

3.1 Harrisian Personhood

In his writings on personhood, Harris[24] often quotes the following passage from John Locke's (1690) *An Essay Concerning Human Understanding*:

> We must consider what person stands for; which I think is a thinking intelligent being, that has reason and reflection, and can consider itself the same thinking thing, in different times and places; which it does only by that consciousness which is inseparable from thinking that seems to me essential to it; it being impossible for anyone to perceive without perceiving that he does perceive.[25]

In this passage from the *Essay*, Locke lays the foundation of two concepts: personhood and personal identity. While Locke is primarily concerned with personal identity in this section, he clearly enunciates what it is to be a person ("a thinking intelligent being", a conscious being with capacity for reason), and subsequently what is required for the person to have a sustained identity (to "consider itself the same thinking thing, in different times and places", to have self-awareness and the spatiotemporal continuity of memory). For Locke, it was these capacities combined that distinguish morally important beings from all others.

[e] The term 'personhood' is not exclusively used by liberal theorists; again see Eberl (note 6).

The distinction is clearly drawn between the biological category of 'human' and the psychological category of 'person'. All human beings (obviously) fall into the first category, but not all necessarily meet the requirements of the second. Furthermore, the Lockean concept of the person is species-neutral. It does not require a person to also be a biological human — it is a logical possibility that a person could be of any species (as long as the necessary criteria are met), or could possibly even be a mentally high-order machine. For Harris, Locke's definition identifies "a range of capacities as the preconditions for personhood [that are] species-, gender-, race- and organic-life-form-neutral".[26]

Harris considers the question of 'when does life begin to matter morally?' in order to establish the criteria for personhood. What are those features of typical human adults we value as morally important? Harris turns to Locke's concept of the person, arguing that these types of capacities are the ones we recognise: "intelligence, the ability to think and reason, the capacity for reflection, self consciousness, memory and foresight".[27] These are also the capacities one needs to value one's own life. For Harris, the connection between the two is significant. If a being has these capacities even at a most basic level, then she is the kind of being that it would be wrong to kill since she values (or is capable of valuing) her life and of imagining her future. To kill her would be to do her a terrible harm as you would deprive her of these things. "Creatures that cannot value their own existence cannot be wronged in this way, for their death deprives them of nothing they value."[28] The capacity to value in a minimally complex way is like an acid test for moral personhood, and the grounds for not harming an individual.

The summative capacity to value one's own existence is important when Harris addresses the idea of personhood as a threshold concept. Hence, he approaches it only once he has identified the Lockean criteria. Accepting that all these features, powers or capacities admit of degrees, Harris suggests that whether a being possesses all to a higher or lower degree is irrelevant since even the most rudimentary levels of these capacities enables them to value their existence.[29] Nor is it the case that your personhood fluctuates with the presence or absence of these features (for example, one does not cease being a person while enduring a blackout) as it is the capacity that counts. While this has been criticised for seeming to employ a certain form of potential, Harris rightly points out that there is a very clear semantic difference between potential and capacity; potential implies that one might eventually develop certain features or abilities, while capacity implies the actual ability, whether exercised or not.[30]

One would hope that the gender- and race-neutrality of personhood is self-evident, but the other two attributes require more discussion. Harris points out that we ascribe personhood status to non-humans such as gods, angels, demons etc.[31] This may or may not be appropriate, but it is indeed commonplace. There is also much merit to the argument against speciesism; it is unlikely that we are the only beings with high-order capabilities in the universe, and it is most likely that we are the only humans. However, the criteria might well be species-specific as it is human beings that have evolved in this particular way over the millennia, and now it is human beings who specify what is valuable and morally important. It would be arrogant of us to presume that other complex and rational beings have evolved similarly to us or that they necessarily value the same features as morally important, particularly when agreement amongst human persons (of Harris' sort) seems ever elusive.

3.2 Criticism of the Personhood View

The Harrisian personhood account of moral status is, philosophically speaking, uncharacteristically straight-forward: "locating the wrongness of killing in one and only one consideration".[32] The personhood approach seems too simplified for such a complex matter as the destruction of human entities. Holm believes that the personhood arguments in favour of embryonic stem cell research constitute a reduction that "proves far too much".[33] Holm's sketch of the personhood position was quoted at the beginning of this section; the position states that analysis of the characteristics of embryos shows they cannot be persons and, therefore, have no moral status, thus permitting us to kill embryos. Holm points out that the personhood account allows for the (non-painful) killing of infants for stem cells as well as the use of embryos for research.

Of course, the permissibility of infanticide is undoubtedly an implication of Harris' position, but there are a few practical reasons why it is unlikely to be pursued.[f] More importantly, for this to be a valid moral objection, it requires some moral principle or theory that defends the notion that infanticide is morally wrong. It cannot be assumed that it is morally wrong simply because it is not common practice, nor is it the case that infanticide is morally wrong because most people believe newborn infants to be special (unless, again, there is an argument to support such a claim). The mere fact that infanticide is permissible is not — alone — sufficient to discredit the entire personhood account. It should also be emphasised here, that it is *permissible* by implication, not advised or required. There may be other reasons why killing an infant would be considered morally wrong, it simply cannot be argued that it is a moral wrong to the infant or that it is an intrinsically morally wrong act. Admittedly, however, the personhood account of moral status focuses moral worth narrowly on persons with little consideration of other entities in a zero-sum game of moral value.

A second objection from Holm highlights the underlying utilitarian nature of the personhood account; the personhood account places no restrictions whatsoever on the use of embryonic material. The embryo, not meeting the personhood threshold, has no moral status at all. This implies that destruction of embryos can be justified if the consequent use of stem cells is beneficial in any way, even if this includes frivolous use. It is possible to imagine that Harris, as a Utilitarian, would place moral restrictions on the use of embryonic material on the basis of some other moral principle, such as scarcity of resources, but not on the basis of the embryo's moral value. This would be consistent, but Holm makes a relatively under-developed although valid point about moral value. Perhaps there are defensible reasons for ascribing moral value to entities such as embryos independent of the question of moral status. This fundamentally challenges the personhood approach, which generally neglects moral value for the more

---

[f] The killing of infants for stem cells is unlikely for a number of practical reasons: It would be unnecessary, as stem cells could be extracted without killing them; it would be time-consuming and costly to engage women in gestating infants for this kind of research; more pliant stem cells can be obtained from a six-day embryo than a neonate.

absolute, all-or-nothing concept of moral status. It is unlikely that ascribing a degree of moral value would conflict with the permissibility of embryonic stem cell research — for moral value would not be equal to moral status — but it could certainly direct the moral acceptability of certain types of research and therapy.

One of the major issues with the Personhood View is that few personhood theorists justify the significance assigned to psychological states and/or capacities for personhood, assuming that it is self-evident. Most rational adult beings engaging in this type of discourse (all of whom are presumed to be of the human species) will accept that moral aspects of human thought and behaviour require a basic minimum of psychological/mental competency. However, many critics of the person-centred approaches have taken issue with the 'philosophical leap' made between what is required and what is deemed sufficient. Harris and other personhood theorists appear to value psychological capacities because they are what allow human beings to have a fully worthwhile life. These capacities count morally although they are mental — not moral — characteristics.[34]

Tom L. Beauchamp[35] heavily criticises the personhood approach to moral status, though he favours the term 'moral standing' for reasons that shall become clear later in the chapter. Beauchamp refers to moral personhood because the theories within this approach assign moral status on the basis of personhood. However, what constitutes a person is a metaphysical matter and as such should be a morally neutral enquiry.[36] Beauchamp does draw a distinction between metaphysical and moral concepts of persons. Metaphysical personhood involves a set of "person-distinguishing psychological properties such as intentionality, self-consciousness, free will, language acquisition, pain reception, and emotion".[37] Ideally, this set of properties details features that are both necessary and sufficient for one to be classified as a person. Moral personhood is quite a different matter, requiring certain moral attributes such as moral agency and moral motivation.[38] It is theoretically possible of course, that one could satisfy the criteria for metaphysical personhood and fail to meet those required by moral personhood. The various personhood theorists fail to address this issue. Although this is a very reasonable objection to the personhood approach, it does not appear to work in favour of its usual critics (the conservative-restrictive theorists) as Beauchamp's concept of moral personhood is undoubtedly much narrower than the personhood described by Harris and others, such as Michael Tooley[39] and Peter Singer.[40]

Beauchamp does not accept that self-consciousness or any other high-order psychological capacity qualifies an entity as a moral person, nor does it bestow any moral standing or moral rights.[41] He does acknowledge that some conditions of metaphysical personhood may be *necessary* but is careful to affirm that none are sufficient. In discussing metaphysical personhood, Beauchamp lists what he sees as the central five characteristics cited by personhood theorists: (i) self-consciousness, (ii) capacity to act on reasons, (iii) capacity to communicate through language systems, (iv) capacity to act freely, and (v) rationality.[42] Proponents of the personhood approach might dispute this list, which serves as a reminder that there is no common list and that there is disagreement as to whether or not one or a set of features is necessary (and sufficient).[43] This ambiguity within the personhood approach also compounds the confusion of a metaphysical–moral connection that remains unexplained. Cognitive abilities alone have no moral importance; it is only when paired with an additional moral theory or moral principle that the metaphysical person can claim to have some sort of moral standing.[44] Of course, Beauchamp insists, an independent moral theory would require independent justification. To illustrate this point, Beauchamp sketches a metaphysical person: "rational, acts purposively, and is self-conscious".[45] There is a possibility that this metaphysical person is also a moral person, but this *metaphysical* description does not preclude the entity from being a sophisticated computer, a dangerous criminal or an 'evil demon'. Metaphysical personhood fails to establish grounds for moral standing since there is no *intrinsic* connection to moral features such as moral agency, moral accountability or moral motivation.

4. THE INTERMEDIATE APPROACHES TO MORAL STATUS

There are two main types of intermediate approach to the question of moral status. The first, the gradualist view, ascribes increasing moral status according to developmental stages of the human organism. The second type of intermediate approach, the moderate view, does not admit of degrees but considers a range of moral features even if they do not amount to personhood.

4.1 The Gradualist View

4.1.1 *Gradualism in Law*

A gradual approach is adopted in law in most countries where abortion is legalised. Both the United Kingdom and the United States (federal law) have legislation that takes a developmental view of personhood. In *Roe v Wade*,[g,46] viability was established as the stage at which legal protection of the foetus comes into force based on a developmental view of interests (in the context of maternal–foetal interests and the interests of the State). The approach of the Human Fertilisation and Embryology Act 1990 (as amended 2008) reflects a gradualist account with regard to embryo research. The law prohibits research on an embryo beyond Day 14.[47] This may appear inconsistent with the Abortion Act 1967, which permits termination up to 24 weeks on social grounds, but it is not necessarily so. The policy we may have regarding research on human organisms may take a different cut-off point to legislation on termination of pregnancy in which the human organism is dependent on another. It is important to note that both sets of legislation adopt a gradualist view of moral worth, but the context of the procedures (research and abortion) requires additional and different ethical considerations to be taken into account.

While the law often reflects social morality, and the two are often interconnected on issues like abortion, the law ascribes *legal* status or legal personhood rather than moral status, even if the latter is a consideration in drafting the legislation.[h]

---

[g] For an overview of the status of the foetus in *Roe v Wade*, see Patricia King (note 47).

[h] See Sheelagh McGuinness in Chapter 3 of this volume for more on the legal construction of the embryo.

4.1.2 *The Gradualist View of Moral Status*

The gradualist approach to moral status recognises the continuity of development in the human organism that eventually gives rise to an adult — one that would typically possess the sort of complex cognitive abilities recognised as morally valuable. Since assigning moral status at any stage of development would be arbitrary, the gradualist suggests that status be incrementally ascribed as the features typical of complex human adults begin to emerge. This means that the more developed an embryo, foetus or newborn is, the greater the moral status we should ascribe to it.[i,48] The most general difficulty with the gradualist approach is that it is difficult to identify a point at which an entity starts to matter morally enough to warrant serious consideration. We need to know when full moral status begins to emerge on the continuum of moral value (from none to full moral status) in order to decide what kind of treatment is acceptable.

4.1.3 *Sumner's Sentience-based Gradualist Account*

L.W. Sumner[49] gives a very thorough gradualist account of moral status. He discusses abortion and the foetus in terms of moral standing, which he sees as equivalent to moral considerability.[50] Sumner holds that moral status is a continuum from none to full status,[51] with the two more established approaches (conservative and liberal, or conception and personhood views as representative of them) at either extreme of the continuum.[52] Sumner finds fault with both for two reasons. First, identifying precise points as the threshold for moral standing requires acceptance of the view that one's moral status abruptly changes from none to full.[53] This Sumner finds implausible, partly because of the other reason. Second, both approaches ascribe status to the foetus uniformly (it always has none or it always has full moral status) despite differences between the early embryo and the late-term foetus.[54]

---

[i] Margaret Olivia Little outlines an interesting and nuanced gradualist position in her paper (see note 48).

Sumner makes his argument for gradual moral standing based on his analysis of possible criteria for the ascription of moral status. He addresses the four main suggestions from across the spectrum of views on moral status: intrinsic value, life, sentience and rationality.[55] Of the four, intrinsic value is the only criterion that is not empirically verifiable.[56] Sumner contends that because of this, one cannot rely on intrinsic value without a theory of intrinsic value. Such a theory would need to provide a basis for intrinsic value. A criterion of intrinsic value would identify the natural property that gives rise to intrinsic worth. Thus we can ascribe moral standing on the basis of the relevant natural property (or properties) without the 'middle man' of intrinsic value. Nothing is gained by referring to intrinsic value. Another unavoidable problem with a criterion of intrinsic value is that it could easily be a matter of convenient description: Attributing intrinsic value to an entity is merely another way of saying that it has moral standing.[57]

The other three criteria (life, sentience and rationality) are all based on identifiable and verifiable states or capacities. Rational beings are a subset of sentient beings, which are a subset of living beings. Sumner claims that the life criterion is the broadest and weakest (most open to challenge) and that rationality is the narrowest and strongest (most likely to be sufficient for moral standing).[58] Rationality is the most difficult of all three to identify; the boundary between non-rational and rational is a complex one to locate because rationality tends to comprise of a set of complex cognitive capacities, such as memory, self-consciousness, reasoning, deliberation, and so on.[59]

While most will accept that rationality is sufficient for moral standing, Sumner argues that it is not necessary.[60] To illustrate this, Sumner refers to the treatment of animals. While many of us eat animals, few among us would be happy if they were ill-treated and slaughtered in an unnecessarily painful manner. This indicates an intuitive ascription of moral value to animals. The greater the number of animals we exclude, based on our chosen criterion of moral standing, the greater the number of human beings we are likely to exclude.[61] Many human beings with limited cognitive function are nevertheless sentient and capable of enjoying their lives, if only in a basic way.[62] While killing them may not be the same as killing a paradigm human being (with full cognitive capacities), Sumner argues that killing them is not morally innocent. He contends that if a being is capable of valuing their own life, then their death costs them something. In this sense, Sumner's conception of valuing one's life is more basic than Harris'; involving enjoyment but not self-reflection.

Being alive is necessary but not sufficient for moral standing. The criterion of rationality though is too strong, according to Sumner.[63] While it is sufficient, it is not a necessary condition for moral standing. It is unsatisfactory because it does not allow in any way for consideration of most morally considerable non-human beings and many human beings, including infants and cognitively compromised adults. Sumner argues that sentience is a defensible alternative. Sentience, like rationality and unlike life, admits of degrees.[64] Consciousness is sufficient for feeling pleasure and pain, the most basic level of sentience, but not for more sophisticated levels of sentience.[65] Interests reflect the higher levels of sentience and it is this that indicates one has moral standing in one's own right.[66] This sentience-based gradualist account of moral status allows for moral consideration of higher order mammals and cetaceans, like chimpanzees and dolphins.[67] It also allows for non-carbon based beings to be included (if encountered), being species-neutral.

Ultimately, Sumner is focused on a morally justifiable policy on abortion. The first trimester foetus is pre-sentient.[68] Early abortions are akin to contraception since abortion on pre-sentient embryos/foetuses is morally innocent. The third trimester foetus, on the other hand, is probably minimally sentient. Abortions at this stage can be considered an 'interpersonal' matter since the foetus has moral standing and should be assessed on a case-by-case basis.[69] Serious justification would be required, such as a threat to the life or health of the pregnant woman.[70] The threshold stage must occur sometime in the second trimester. The moral standing of the foetus at this stage is indeterminate, and we cannot know if a liberal approach is appropriate or if a conservative approach is required.[71] Sumner suggests that as long as this indeterminacy persists (scientific knowledge may become more precise), it is best to have an abortion as early as possible.[72] He does not invoke a precautionary principle and advise the stricter justification for abortion at this stage. Sumner suggests that the lack of clarity, which is tolerable and desirable where exactitude is impossible, can be resolved pragmatically in policy by drawing the cut-off point midway through the second trimester.[73] This would allow for embryo and foetal experimentation (at an early stage).

4.1.4 *Lockwood's Brain Birth Account*

Some suggest intermediate stages (between conception and full rationality/personhood) for the beginning of moral status. Michael Lockwood[74,75] introduced the concept of 'brain birth' as the beginning of a *morally* important life.[j] Taking the medical and social acceptance of brain death as the end of a moral life, Lockwood suggests that the initiation of brain development would be the appropriate stage for the beginning of a morally valuable life. Lockwood wishes to trace back to the beginning of the brain's continuous existence, because just as we cease to exist when the relevant parts of the brain cease functioning, we begin as soon as the brain begins functioning.[76] This seems somewhat imprecise given that we cannot know how developed the brain needs to be so that significant parts of the brain are fully functioning. The system of the brain develops gradually, but Lockwood ascribes status from the beginning of development not at any particular stage of significance. There is a *prima facie* rule against harming or killing any entity with a living human brain (no matter how rudimentary), but the justifications may vary according to the level of development. Hence, the early embryo has no moral protection, and the justifications for foetal harm must be increasingly serious as the brain and neurological systems develop.

[j] Lockwood's 'brain birth' view falls uneasily between a gradualist and non-gradualist moderate account of moral status. On the one hand, it acknowledges gradual development, but on the other, he establishes a cut-off of 'brain birth' based on the notion that it is the beginning of the essentially rational human beings we become (fitting broadly with rationality/personhood views of moral status).

4.1.5 *Objections to the Gradualist View*

Harris[77] rejects the gradualist approach. If we can establish which features precisely comprise a normal human adult of personhood (and Harris believes we can), then it may be possible to more accurately determine the stage at which they are present, and the gradualist approach would be unnecessary. However, it is precisely this lack of accuracy in determining the relevant stages that gives rise to the gradualist account of moral status. In addition, a more complex and fluid understanding of moral value can be seen in the gradualist position, where somewhat moral entities have some but not full moral status.

4.2 The Moderate View

4.2.1 *Metaphysical versus Moral Personhood and the Moral Community*

As mentioned in the Personhood section above, Beauchamp has criticised strictly metaphysical personhood accounts of moral status. Beauchamp highlights that most, if not all, the metaphysical properties identified can be present in varying degrees, and argues that placing value on these types of characteristics requires acceptance of differing degrees of personhood for human *and* non-human animals, and also the possibility that some non-human animals may have a higher degree of personhood than some human persons.[78] While some personhood theorists (certainly Harris and Singer) apply personhood as a threshold concept, they are more than prepared to ascribe higher moral value to some non-human animals, such as primates, than they would to some human animals, such as late-term foetuses, who have no moral status but whose sentience is morally considerable.

Whereas metaphysical personhood has many apparent difficulties, Beauchamp sees moral personhood as "relatively uncomplicated".[79] A being is a moral person if it is capable of morally evaluating actions as right and wrong, and if it has motivations that can be judged morally. Moral personhood is sufficient for moral standing, and moral persons are members of the moral community, hence qualifying for all its rights and benefits as well as the responsibilities.[80] Moral personhood has an interpersonal quality, reciprocity and communal expectations, in particular the recognition and judgement of other moral agents that matters. However, Beauchamp's moral personhood creates an exceedingly narrow category of moral persons.

Beauchamp, however, ascribes moral standing even to non-persons: "certain *noncognitive* and *nonmoral* properties are sufficient to confer a measure of moral standing".[81] Sentience and the capacity for an emotional life qualify any creature for 'moral standing'. Furthermore, the interests (e.g. in avoiding pain) of some non-persons would outweigh certain interests of persons.[82] While some beings may not qualify as moral persons, the moral community has an obligation to protect them, and possibly even to ascribe full moral standing (including moral rights) — though only on the basis of something other than moral personhood.[83] This moderate account of moral standing (rather than moral status) takes the complexities of a full moral community into account in a way that personhood accounts cannot.

4.2.2 *Robertson's Symbolic Value*

While moral status is generally based on intrinsic properties, moral value can be ascribed for a variety of reasons. John Robertson argues that "[o]ne can deny that something has intrinsic value as a moral subject, yet still value it or accord it meaning because of the associations or symbolism that it carries".[84] We give symbolic value to many things and people. Symbolic value can be attributed to entities with no other kind of moral value, to entities that ordinarily would not have moral value, such as the Blarney Stone.[k] While most pieces of bluestone have no moral value, this

[k] The Blarney Stone is a block of bluestone embedded in the walls of Blarney Castle in County Cork, Ireland. It has a number of dubious legends attached to it as to its provenance but appears to have been built into the bombardments several centuries ago. It is famous for endowing those who kiss it (at great risk to their lives) with the 'gift of the gab' (a skill akin to eloquent flattery).

one has a special history and ritual attached. Though a somewhat trivial example, this piece of bluestone does have symbolic value, and it is likely that many would be affected by its loss or destruction. Symbolic value can be attributed to entities with full moral status, for example the Queen of England has moral status as a moral agent (or person) but also has symbolic value as the monarch of England. This warrants special treatment, such as personal security in the Queen's case or weather-protection in the case of the Blarney Stone. The Queen is of far greater moral value than the Blarney Stone because she also has the ultimate moral value of moral status based on intrinsic properties.

Robertson has expounded on the concept of symbolic value as a way of answering the questions often raised about 'special respect' for embryos.[85] Recognising symbolic moral value in the embryo is a way to avoid trivialisation of that which the embryo symbolises, human life. Symbols do not make the same moral claims on us as those with moral status do. Furthermore, they tend to be subjective and personal.[86] If the Blarney Stone were destroyed, it would likely only affect Irish people, or people with an affinity for Ireland and the Blarney Stone. The symbolic value of the Blarney Stone is greater for Irish people than for Ghanaians (few of whom are likely to have heard about or care about the Blarney Stone). This subjectivity is one of the reasons some argue against the concept of symbolic value. It does not seem to be sufficient to warrant much moral consideration on its own and adds little to entities that have stronger moral value based on intrinsic capacities or properties. This, however, is a judgement about the usefulness of the concept, not an evaluation of whether it is a concept that accurately reflects one of the various ways human beings with full moral status tend to ascribe moral value to entities.

Robertson specifically addresses the question of embryonic stem cell research with regard to symbolic value. Because it is an extrinsic valuation, symbolic value does not have the force to prohibit practices, but can limit them. Thus, according to Robertson, symbolic value would permit the use of supernumerary IVF embryos in research but only for morally worthy reasons, such as medical research, and not for frivolous reasons, such as cosmetics research. As Robertson describes it, the evaluation of what can acceptably be done to an entity with symbolic value is something like a 'symbolic costs–moral benefits' analysis.[87] It is a relative judgement, weighing up the benefits and losses of sacrificing an entity with symbolic value. The more important the benefits, the easier it is to justify harmful treatment of an entity with symbolic value (but here it would be important to note that it would be entities with *only* symbolic value).

4.2.3 *Warren's Multi-criterial View*

Mary Anne Warren's[88] multi-criterial theory of moral status is both more permissive than the conception view and more restrictive than the personhood view; it is both more demanding and less demanding. It combines intrinsic properties with extrinsic considerations to provide seven core principles of moral status. Superficially, it looks like a patch-work, catch-all theory that takes the important elements of all theories and combines them. It is more sophisticated than this, but it does look to the notable elements of other inadequate or flawed approaches to moral status in order to develop a complex understanding of what is involved in moral status. This multi-criterial approach also takes account of the real context in which we live.

Warren says "social and biological relationships shape our moral obligations towards many entities".[89] This is a perfectly reasonable assertion; however it concerns our duties to entities, not the *value* of those entities. Taking social and biological relationships into account is a contingent aspect of decision-making. Like Robertson's symbolic value, it is not ultimately determinative regarding the moral value of an entity. Warren's seven principles of moral status are: the respect for life principle, the anti-cruelty principle, the agent's rights principle, the human rights principle, the ecological principle, the inter-specific principle and the transitivity of respect principle.[90] Some of these principles are tied to intrinsic properties, such as the anti-cruelty principle, which is based on the intrinsic capacity for sentience. Others are related to extrinsic properties, such as the transitivity of respect principle, which instructs us to respect other moral agents' attributions of moral status "to the extent that is feasible and morally permissible"[91] and only as long as respecting these attributions is congruent with the other six principles.

The strength of the multi-criterial approach is that it acknowledges serious obligations even to entities without full moral status. The use of the term moral status, albeit on a sliding scale, would lend normative force to the notion that other entities, such as non-human animals, might warrant moral consideration.

5. MORAL STATUS, MORAL VALUE

Neither of the absolute accounts of moral status is satisfactory. The conservative–restrictive accounts rely too heavily on unsubstantiated beliefs, unjustifiable principles and ambiguous concepts, such as human dignity and sanctity. The liberal–permissive accounts, based largely on personhood, fail to provide adequate justification for relying on psychological criteria to indicate moral status, and many of the personhood theories are very narrow in focus, seeming to give little moral value to anything other than persons. The 'third way' of the intermediate approaches is not entirely satisfactory either; symbolic value, as an extrinsic quality, is highly variable, and gradualist accounts allow for the ascription of moral status to virtually any type of entity. While the gradualist accounts are useful for the more comprehensive view of moral value they encapsulate, the use of moral status in degrees rather than as a threshold is problematic.

5.1 The Concept of Moral Status

The concept of moral status appears to have a number of core features, although not all approaches use a concept of moral status that includes all of them. Two of the most common are inviolability and threshold. The nature of moral status as a threshold concept is something that is challenged by all approaches; some seem too absolute while the gradualist accounts raise other difficulties. On the one hand, gradualist approaches inevitably generate indefinite lower limits. On the other hand, an absolute threshold for moral status on personhood accounts excludes too many (in particular those entities very close to the threshold) and may include too many on the conservative view.

If we want to associate moral status with inviolability, it would appear that it must be a threshold. If we are not concerned about inviolability, then a special term like 'moral status' seems pointless. Thus it appears that moral status, to be conceptually useful, is necessarily a threshold concept. Gradualists appear to use moral status as an indicator of serious moral value where minimal moral status does not confer inviolability but considerability. In such cases, it would perhaps be more meaningful to speak in terms of serious moral value and moral considerability. Those entities with minimal moral status tend to require only minimal justification when inflicting harm, and there is no reason why moral value cannot impose the same requirements and *prima facie* duties. The importance of the gradualist concept of moral status is that it forces us to consider more seriously those entities with minimal moral status, or what non-gradualists might class as moral worth, moral considerability or moral standing.

5.2 A Sliding Scale of Moral Value

There could still be a sliding scale, similar to that envisaged by many gradualists; it would just begin with minimal moral value and continue to (full[1]) moral status at the peak of the scale. This scale may take significant (necessary but not sufficient) features into account; there is certainly support for the argument that sentience ought to be taken far more seriously, for example. This would have implications for foetuses, general treatment of non-human animals and the use of non-human animals in research, but would not apply to the non-sentient early embryo.

The significant point here is that the scale concerns moral value rather than status, and that status continues to have a distinct and ultimate

[1] Of course, 'full' would be redundant on this sliding scale of moral value because there would only be an absolute moral status.

meaning. Moral status would require moral capacities, such as moral integrity, for which certain psychological capacities would be necessary but not sufficient. This would exclude a high percentage of known entities, including some human beings. However, extrinsic value, such as relational and social value, would still count. Greater significance would be ascribed to entities with morally considerable capacities, such as sentience, also. Similar to gradualist accounts of moral status, increasingly serious and significant justifications need to be provided for inflicting harm or death the further up the scale of moral status an entity is located. Most importantly, moral status would not be the only criterion for serious moral consideration; to put it conversely, lack of moral status would not preclude an entity from serious moral consideration in decisions affecting them. This somewhat modified version of Warren's account of moral status would more accurately be described as a multi-criterial account of moral considerability, whereby entities are morally considerable whether that is due to moral value or moral status.

5.3 The Moral Value of the Embryo

It appears, from a critical survey of the literature, that absolute protection of the early human embryo cannot be morally justified on the basis of its moral status. Certain human embryos may have relational value (derived from the value placed on it by entities with moral status) and/or symbolic value (ascribed by entities with moral status). Whether the embryo has intrinsic moral value is difficult to determine. If it does, it could not be sufficient to prohibit all experimentation. It may prohibit certain types of experimentation (so-called trivial research), but not all.

6. THE MORAL STATUS OF THE EMBRYO AND PUBLIC POLICY

What are the implications of this for stem cell policy? Only that research on human embryos cannot be prohibited based on their moral status. As mentioned above (Section 5.2), there may be other reasons to give embryos serious consideration. This might include the embryos relational or symbolic value. The social value of the embryo might also be significant (depending on the society). That is to say, the high social value ascribed to embryos (even without philosophical foundation) might warrant greater consideration because of the implications destructive research would have for that community.

Embryonic stem cell research, like abortion, involves inherently personal moral beliefs regarding the nature of the human embryo. These are not always philosophically sound, and it is unlikely that 'public opinion' will ever be entirely based on philosophically valid arguments. However, public policy is not about enforcing private morality; rather, it is concerned with "providing a framework of peace and order within which people may exercise their personal liberty".[92] Public policy reasoning ought not to privilege particular religious or philosophical views. However, proposing a policy that ignores a significant section of the public because the position has a particular theological or philosophical basis could be equally problematic in its divisiveness. Bearing this in mind, it would seem that a broader, more holistic account of moral value (rather than a narrow moral status account of the Conception or Personhood sort) would be helpful in providing a common ground and language for discussion. Such an account does not immediately exclude or include marginal entities but provides, at least, a moral context for debate.

REFERENCES

1. Holland S. (2001) Beyond the Embryo: A Feminist Appraisal of the Embryonic Stem Cell Debate. In: Holland S, Lebacqz K, Zoloth L (eds.), *The Human Embryonic Stem Cell Debate: Science, Ethics and Public Policy.* MIT Press, Cambridge, MA, pp. 73–86.
2. Warren MA. (2005) *Moral Status: Obligations to Persons and Other Living Things.* Oxford University Press, Oxford.
3. Smith SW. (2008) Precautionary Reasoning in Determining Moral Worth. In: Freeman MDA (ed.), *Law and Bioethics.* Oxford University Press, Oxford, pp. 197–212.

4. Conley JJ. (2002) Delayed Animation: An Ambiguity and Its Abuses. In: JW Koterski (ed.), *Life and Learning* XII, pp. 159–168, p. 159. Available at www.uffl.org/vol12/conley12.pdf (Accessed 16 June 2011).
5. Murray D. (2007) *What Is a Person?* Veritas, Dublin, pp. 20–22.
6. Eberl JT. (2000) The Beginning of Personhood: A Thomistic Biological Analysis. *Bioethics* **14(2)**: 134–157.
7. Pope John Paul II. (1995) Evangelium Vitae. Papal Encyclical, §60. Available from http://www.vatican.va/edocs/ENG0141/_INDEX.HTM (Accessed 16 June 2011).
8. Macklin R. (2003) Dignity Is a Useless Concept. *British Medical Journal* **327:** 1419–1420.
9. Macklin R. (2003) Dignity Is a Useless Concept. *British Medical Journal* **327:** 1419–1420, p. 1419
10. Ashcroft RE. (2005) Making Sense of Dignity. *Journal of Medical Ethics* **31:** 679–682.
11. Schroeder D. (2008) Dignity: Two Riddles and Four Concepts. *Cambridge Quarterly of Health Care Ethics* **17:** 230–238.
12. Kant I. (1785/1997) *Groundwork of the Metaphysics of Morals*. Gregor MJ (trans). Cambridge University Press, New York, NY, p. 429 (Akadamie pagination).
13. Kain P. (2009) Kant's Defense of Human Moral Status. *Journal of the History of Philosophy* **47(1):** 59–102, p. 61.
14. Alvarez Manninen B. (2008) Are Human Embryos Kantian Persons?: Kantian Considerations in Favor of Embryonic Stem Cell Research. *Philosophy, Ethics, and Humanities in Medicine* **3(4):** 3.
15. Beyleveld D, Brownsword R. (2001) *Human Dignity in Bioethics and Biolaw*. Oxford University Press, Oxford — 2004, Reprint edition, pp. 53–54.
16. Alvarez Manninen B. (2008) Are Human Embryos Kantian Persons?: Kantian Considerations in Favor of Embryonic Stem Cell Research. *Philosophy, Ethics, and Humanities in Medicine* **3(4):** 3 (emphasis in original).
17. Beyleveld D, Brownsword R. (2001) *Human Dignity in Bioethics and Biolaw*. Oxford University Press, Oxford — 2004, Reprint edition.
18. Beyleveld D, Brownsword R. (2001) *Human Dignity in Bioethics and Biolaw*. Oxford University Press, Oxford — 2004, Reprint edition, p. 116.
19. Beyleveld D, Brownsword R. (2001) *Human Dignity in Bioethics and Biolaw*. Oxford University Press, Oxford — 2004, Reprint edition, pp. 106–107.
20. Beyleveld D, Brownsword R. (2001) *Human Dignity in Bioethics and Biolaw*. Oxford University Press, Oxford — 2004, Reprint edition, p. 125.

21. Beyleveld D, Brownsword R. (2001) *Human Dignity in Bioethics and Biolaw.* Oxford University Press, Oxford — 2004, Reprint edition, p. 125.
22. Beyleveld D. (2000) The Moral Status of the Human Embryo and Fetus. In: Haker H, Beyleveld D (eds.), *The Ethics of Genetics in Human Procreation.* Ashgate, Aldershot, pp. 59–86, pp. 59–60.
23. Holm S. (2003) The Ethical Case against Stem Cell Research. *Cambridge Quarterly of Health Care Ethics* **12:** 372–383, p. 372.
24. Harris J. (1985) *The Value of Life: An Introduction to Medical Ethics.* Routledge & Kegan Paul, London, p. 15.
25. Locke J. (1690/1999) *An Essay Concerning Human Understanding.* Ch.XXVII, Bk. II: 217. Available at http://www2.hn.psu.edu/faculty/jmanis/locke.htm (Accessed 16 June 2011).
26. Harris J. (1999) The Concept of the Person and the Value of Life. *Kennedy Institute of Ethics Journal* **9(4):** 293–308, p. 303.
27. Harris J. (1999) The Concept of the Person and the Value of Life. *Kennedy Institute of Ethics Journal* **9(4):** 293–308, p. 303.
28. Harris J. (1985) *The Value of Life: An Introduction to Medical Ethics.* Routledge & Kegan Paul, London, p. 19.
29. Harris J. (1999) The Concept of the Person and the Value of Life. *Kennedy Institute of Ethics Journal* **9(4):** 293–308, p. 303.
30. Harris J. (1985) *The Value of Life: An Introduction to Medical Ethics.* Routledge & Kegan Paul, London, pp. 25–26.
31. Harris J. (1999) The Concept of the Person and the Value of Life. *Kennedy Institute of Ethics Journal* **9(4):** 293–308, p. 293.
32. Holm S. (2003) The Ethical Case against Stem Cell Research. *Cambridge Quarterly of Healthcare Ethics* **12:** 372–383, p. 373.
33. Holm S. (2003) The Ethical Case against Stem Cell Research. *Cambridge Quarterly of Healthcare Ethics* **12:** 372–383, p. 372.
34. Marquis D. (1989) Why Abortion is Immoral. *Journal of Philosophy* **86(4):** 183–202, p. 186
35. Beauchamp TL. (1999) The Failure of Theories of Personhood. *Kennedy Institute of Ethics Journal* **9(4):** 309–324.
36. Beauchamp TL. (1999) The Failure of Theories of Personhood. *Kennedy Institute of Ethics Journal* **9(4):** 309–324, p. 309.
37. Beauchamp TL. (1999) The Failure of Theories of Personhood. *Kennedy Institute of Ethics Journal* **9(4):** 309–324, pp. 309–310.
38. Beauchamp TL. (1999) The Failure of Theories of Personhood. *Kennedy Institute of Ethics Journal* **9(4):** 309–324, p. 310.
39. Tooley M. (1983) *Abortion and Infanticide.* Clarendon Press, Oxford.

40. Singer P. (1993) *Practical Ethics.* Cambridge University Press, Cambridge.
41. Beauchamp TL. (1999) The Failure of Theories of Personhood. *Kennedy Institute of Ethics Journal* **9(4):** 309–324, pp. 310–311.
42. Beauchamp TL. (1999) The Failure of Theories of Personhood. *Kennedy Institute of Ethics Journal* **9(4):** 309–324, p. 311.
43. Beauchamp TL. (1999) The Failure of Theories of Personhood. *Kennedy Institute of Ethics Journal* **9(4):** 309–324, p. 312.
44. Beauchamp TL. (1999) The Failure of Theories of Personhood. *Kennedy Institute of Ethics Journal* **9(4):** 309–324, p. 312.
45. Beauchamp TL. (1999) The Failure of Theories of Personhood. *Kennedy Institute of Ethics Journal* **9(4):** 309–324, p. 312.
46. King P. (1978–79) The Juridical Status of the Fetus: A Proposal For Legal Protection of the Unborn. *Michigan Law Review* **77:** 1647–1687.
47. Human Fertilisation and Embryology Act 1990 (as amended 2008), ss. 3–4.
48. Little MO. (2008) Abortion and the Margins of Personhood. *Rutgers Law Journal* **39:** 331–348.
49. Sumner LW. (1981) *Abortion and Moral Theory.* Princeton University Press, Princeton, NJ.
50. Sumner LW. (1981) *Abortion and Moral Theory.* Princeton University Press, Princeton, NJ, p. 26.
51. Sumner LW. (1981) *Abortion and Moral Theory.* Princeton University Press, Princeton, NJ, p. 26.
52. Sumner LW. (1981) *Abortion and Moral Theory.* Princeton University Press, Princeton, NJ, p.124.
53. Sumner LW. (1981) *Abortion and Moral Theory.* Princeton University Press, Princeton, NJ, p. 125.
54. Sumner LW. (1981) *Abortion and Moral Theory.* Princeton University Press, Princeton, NJ, p. 125.
55. Sumner LW. (1981) *Abortion and Moral Theory.* Princeton University Press, Princeton, NJ, p.129.
56. Sumner LW. (1981) *Abortion and Moral Theory.* Princeton University Press, Princeton, NJ, p. 131.
57. Sumner LW. (1981) *Abortion and Moral Theory.* Princeton University Press, Princeton, NJ, p. 131.
58. Sumner LW. (1981) *Abortion and Moral Theory.* Princeton University Press, Princeton, NJ, pp. 131–132.
59. Sumner LW. (1981) *Abortion and Moral Theory.* Princeton University Press, Princeton, NJ, p. 137.

60. Sumner LW. (1981) *Abortion and Moral Theory.* Princeton University Press, Princeton, NJ, p. 138.
61. Sumner LW. (1981) *Abortion and Moral Theory.* Princeton University Press, Princeton, NJ, p. 139.
62. Sumner LW. (1981) *Abortion and Moral Theory.* Princeton University Press, Princeton, NJ, p. 140.
63. Sumner LW. (1981) *Abortion and Moral Theory.* Princeton University Press, Princeton, NJ, p. 142.
64. Sumner LW. (1981) *Abortion and Moral Theory.* Princeton University Press, Princeton, NJ, p. 143.
65. Sumner LW. (1981) *Abortion and Moral Theory.* Princeton University Press, Princeton, NJ, p. 142.
66. Sumner LW. (1981) *Abortion and Moral Theory.* Princeton University Press, Princeton, NJ, p. 145.
67. Sumner LW. (1981) *Abortion and Moral Theory.* Princeton University Press, Princeton, NJ, p. 144.
68. Sumner LW. (1981) *Abortion and Moral Theory.* Princeton University Press, Princeton, NJ, p. 149.
69. Sumner LW. (1981) *Abortion and Moral Theory.* Princeton University Press, Princeton, NJ, p. 126.
70. Sumner LW. (1981) *Abortion and Moral Theory.* Princeton University Press, Princeton, NJ, p. 152.
71. Sumner LW. (1981) *Abortion and Moral Theory.* Princeton University Press, Princeton, NJ, p. 153.
72. Sumner LW. (1981) *Abortion and Moral Theory.* Princeton University Press, Princeton, NJ, p. 153.
73. Sumner LW. (1981) *Abortion and Moral Theory.* Princeton University Press, Princeton, NJ, p. 158.
74. Lockwood M. (1985) *Moral Dilemmas in Modern Medicine.* Oxford University Press, Oxford.
75. Lockwood M. (2001) The Moral Status of the Human Embryo. *Human Fertility* **4(4)**: 267–269.
76. Lockwood M. (1985) *Moral Dilemmas in Modern Medicine.* Oxford University Press, Oxford, p. 23.
77. Harris, J. (1999) The Concept of the Person and the Value of Life. *Kennedy Institute of Ethics Journal* **9(4)**: 293–308.
78. Beauchamp TL. (1999) The Failure of Theories of Personhood. *Kennedy Institute of Ethics Journal* **9(4)**: 309–324, p. 314.

79. Beauchamp TL. (1999) The Failure of Theories of Personhood. *Kennedy Institute of Ethics Journal* **9(4)**: 309–324, p. 315.
80. Beauchamp TL. (1999) The Failure of Theories of Personhood. *Kennedy Institute of Ethics Journal* **9(4)**: 309–324, p. 315.
81. Beauchamp TL. (1999) The Failure of Theories of Personhood. *Kennedy Institute of Ethics Journal* **9(4)**: 309–324, p. 316 (emphasis in the original).
82. Beauchamp TL. (1999) The Failure of Theories of Personhood. *Kennedy Institute of Ethics Journal* **9(4)**: 309–324, p. 317.
83. Beauchamp TL. (1999) The Failure of Theories of Personhood. *Kennedy Institute of Ethics Journal* **9(4)**: 309–324, p. 316.
84. Robertson JA. (1999) Ethics and Policy in Embryonic Stem Cell Research. *Kennedy Institute of Ethics Journal* **9(2)**: 109–136, p. 118.
85. Robertson JA. (1995) Symbolic Issues in Embryo Research. *Hastings Center* Report **25(1)**: 37–38, p. 37.
86. Robertson JA. (1995) Symbolic Issues in Embryo Research. *Hastings Center* Report **25(1)**: 37–38, p. 38.
87. Robertson JA. (1999) Ethics and Policy in Embryonic Stem Cell Research. *Kennedy Institute of Ethics Journal* **9(2)**: 109–136, p. 127.
88. Warren MA. (2005) *Moral Status: Obligations to Persons and Other Living Things.* Oxford University Press, Oxford.
89. Warren MA. (2005) *Moral Status: Obligations to Persons and Other Living Things.* Oxford University Press, Oxford, p. 146.
90. Warren MA. (2005) *Moral Status: Obligations to Persons and Other Living Things.* Oxford University Press, Oxford, pp. 148–172.
91. Warren MA. (2005) *Moral Status: Obligations to Persons and Other Living Things.* Oxford University Press, Oxford, p. 170.
92. Green RM (ed.). (2000) *Encyclopedia of Ethical, Legal and Policy Issues in Bio-technology* (Vol. 2). Wiley-Interscience, New York, Charlesworth cited at p. 636.

3

The Construction of the Embryo and Implications for Law

Sheelagh McGuinness[a]

The purpose of this chapter is to make the case for the embryo as a constructed concept. I discuss some different constructions of the embryo and show how they carry in them implicit accounts of the values of the embryo. Towards the end of the chapter, I briefly describe the legal implications of different constructions and show how in different contexts the embryo may have different values. I use the recent decision of the Advocate General in the European Court of Justice in Oliver Brüstle v Greenpeace as a case study to highlight Franklin's description of the embryo as a legally indeterminate appellation. By examining the way in which the embryo is constructed, we can come to a clearer idea of what aspects of the construction are important to different groups. That is not to say that taking a constructionist approach will overcome the differences in approach to the embryo — rather it will enable us to better understand the disagreements and move policy forward in a way that can better accommodate these differing views.

Centre for Professional Ethics, School of Law, Keele University, Staffordshire, ST5 5BG, United Kingdom. Email: s.mcguinness@keele.ac.uk

[a] I would like to thank Marie Fox, Marie Jacob and Sorcha Uí Chonnachtaigh for comments on a previous draft of this chapter. Thank you also to John Coggon.

> 'Then you should say what you mean,' the March Hare went on. 'I do,' Alice hastily replied; 'at least — at least I mean what I say — that's the same thing, you know.' 'Not the same thing a bit!' said the Hatter. 'Why, you might just as well say that "I see what I eat" is the same thing as "I eat what I see!"[1]

1. INTRODUCTION

Many terms used in bioethical debate are of sufficiently varied or ambiguous meaning to make it worthwhile to subject their use to close scrutiny to avoid the possibility that people are talking past each other — examples include 'person', 'human being', 'dignity' or 'humanity'. Specifically in relation to the issues addressed in this book, the word 'embryo' is itself subject to much manipulation and variation in meaning that will be teased out throughout this chapter. Michael Tooley, speaking in relation to the use of the term 'human being' in the abortion debate, summed up the linguistic hurdles that must be overcome as follows:

> When this ambiguity goes unnoticed, as it usually does, the result is a discussion in which people talk past one another, generating considerable heat, but very little illumination. In approaching moral questions philosophically it is important to be sensitive to such ambiguity, and to attempt to make explicit the different possible meanings of crucial terms. When this is done, fruitful discussion becomes possible.[2]

The embryo is a strong yet porous concept. It is strong in the impact that it has on the imagination but porous in the vagaries of exactly what the concept consists in. As Franklin describes:

> As has become clear as *in vitro* human embryos have acquired an increasingly public civic life over the past half century, over-attachment to an idealized and singular invocation of 'the' embryo is both partisan and illogical. It is an ethical framing device that human embryos, in their enormous diversity, by definition exceed and overflow. Scientifically, 'embryo' is a basket category — like 'clone', it is famously imprecise. Legally, it is an equally indeterminate appellation, and philosophically it has been the subject of debate for more than two millennia.[3]

Analysing how people talk about embryos helps to unpack some of the more difficult philosophical questions about the permissibility of different activities. An understanding of what different words, used in different contexts, can mean to different sections of the debate is not only helpful, but necessary, if we wish to move from a debate focused on the extremes. At times, it appears that the language of the extremes dominates the debate; an individual can be pro-life (or anti-life), pro-choice (or anti-choice).

In this chapter, I do not hope to solve the many disputes over the permissibility of activities such as human embryonic stem cell (hESC) research which involve embryos. My aims are much more modest. The purpose of this chapter is to make the case for the embryo as a constructed concept. I will then move to discuss different constructions and show how they carry in them implicit accounts of the values of the embryo. Towards the end of the chapter, I will briefly describe the legal implications of different constructions and show how in different contexts the embryo may have different values. The recent decision of the Advocate General of the Court of Justice in *Oliver Brüstle v Greenpeace*[4] will be used as a case study to highlight Franklin's description of the embryo as a legally indeterminate appellation.

2. WHAT IS CONSTRUCTION?

The first time I taught a session on the Construction of the Embryo was to undergraduate law students on a course titled 'Law, Science, and Society'. I was amused to find out that my students thought they would be attending a class on the various ways in which embryos can be created (in a lab or in a bedroom) and how the law regulates these different *constructions*. That is not the sort of construction I am interested in; although, how the embryo is created can affect the level of legal protection it is afforded. Rather I follow Mulkay and others in thinking that the embryo is a 'socially, politically, and culturally constructed entity', and to this list I want to add legally.[b] It is through our various constructions of the embryo that we come to a view about what it is permissible to do to it and with it.

[b] There is a growing body of literature that holds this view. See notes 5–13.

Different constructions of the embryo have different ethical and legal implications. Jacob and Prainsack summarise the rationale of an approach like this as the "need to analyse embryos in the context of networks of social and biological relations they are embedded in".[14] Kathryn Ehrich *et al.* describe what this analysis means for understanding of the embryo:

> Embryos have been thought of as 'social objects', i.e., objects which are understood to have socially constructed meanings rather than intrinsic natures in a variety of ways, for example, contested political object, commercial entity, sacred object, highly regulated legal object. How embryos are constituted as social objects is subject to continual and contextualized change.[15]

There is much 'embryo talk' in the discussion of how we should regulate assisted reproductive technologies and stem cell research. In analysing this talk, we will begin to see the various ways in which the embryo is constructed in different contexts and also the ethico–legal implications of this. How we construct the embryo affects our view of the permissibility of different activities. There are various ways in which we can understand what this means for our analysis of stem cell research. We could argue, as Devolder and Harris do, that "the embryo is an irredeemably ambiguous entity" and focus our analysis instead on other more concrete concerns.[16] Here, I will take a different approach; rather than taking the ambiguity of the embryo as a conclusion for how it is we understand embryos, I will take this as a starting point for interrogation.

Construction in this context can be understood as the belief that our knowledge is 'constructed', rather than simply objectively knowable.[17] A distinction is made between 'reality' and 'knowledge', the latter being our perception of the former. A constructionist view of the embryo holds that our knowledge of the embryo is a construct that can be subject to interpretation. Constructionism can broadly be either 'hard' or 'soft'. Hard constructionists hold that our account of the embryo is based on construction alone, and this can tell us nothing of what the embryo is really like. Soft constructionism, on the other hand, holds that although we do have differing constructions of what the embryo is, there is a 'real' embryo outside of this, and when we examine a variety of constructs we get a sense of what the embryo is *really* like, to some extent. It is the latter approach that I adopt in this chapter. This means that throughout the chapter, I will distinguish between the embryo as an objective entity and the embryo as a constructed concept. The objective embryo consists of a core of settled meaning that is subject to various interpretations, and it is these interpretations that give rise to various constructions of the embryo. Our knowledge of the embryo can therefore be understood as an interpretation of particular features of the embryo, and this gives rise to different constructions as to what an embryo is.

Discussion of what embryos are is made all the more difficult by the fact that the different terms that can be used to describe this entity are contested, and there is much disagreement concerning the most appropriate terms. Often it is evident that there is a conflation of descriptive and normative accounts of what an embryo is. In discussing the embryo, this conflation of normative and descriptive is particularly powerful when normative force is given to biological or scientific description.[c] The terminology we use to describe the embryo has become highly politicised. This is evidenced in the debate that surrounded the use of the terms 'embryos' and 'pre-embryos'. Lee Silver describes the controversy around the use of the terms as follows:

> It is rare that scientists change well-established terminology and such changes normally occur only in response to new scientific understanding that invalidates the use of earlier terms. Yet our scientific understanding of early embryonic development has remained essentially unchanged for more than half a century. So why is there suddenly a need to adopt a new word? ... The term pre-embryo has been embraced wholeheartedly ... for reasons that are political, not scientific.[19]

At the time that Silver was writing, the term 'pre-embryo' had come to be used to describe embryos at the very earliest stages of existence prior to individuation. Those who used the term often tried to argue that even if

---

[c] See note 18 on the 'emerging pro-science agenda'.

embryo destruction was generally not permissible, the destruction of pre-embryos was acceptable as these 'embryos' were different. It is interesting to note that in the time since Silver wrote this, the term pre-embryo has been largely discarded, especially in the United States (US). Speaking in a United Kingdom (UK) context, Williams *et al.* state:

> Whereas opponents defined the embryo at any stage in terms of its human potential, some proponents implied that, before 14 days, the cells did not constitute an embryo at all. They made efforts to distinguish between 'proper' embryos and things they called the 'early-stage embryo' or 'blastocyst'. In doing so, they were building on pre-existing initiatives in the run up to the Government's Human Fertilization and Embryology Bill when efforts were made to distinguish the 'pre-embryo'.[d,20]

More recently we have seen a further controversy and changing of terminology in the passing through Parliament of the 2008 amendments to the Human Fertilisation and Embryology Act 1990 in the debate over 'hybrid' embryos. Initially, the government had decided that the Act would completely prohibit the creation of animal/human hybrid embryos.[22] However, following strong lobbying from the scientific community, the government made a U-turn, and the wording of the Bill was changed so as to permit the creation of admixed embryos for the purposes of research.[23] Opponents of the creation of these entities felt that the change in terminology was an exercise in linguistic manoeuvring by the government in order to hide the true meaning of the activities involved.[24] These examples serve to illustrate how often what an embryo is described as will be controversial, and also reflect views of what the describer thinks it is permissible to do to it.

As mentioned, this chapter has as its primary focus 'the embryo' as a broad category, and throughout, a distinction will be made between different types of embryo; for example, 'the research embryo' or 'the reproductive embryo'.[e] In making these distinctions, I will characterise

---

[d] For a discussion of the importance of the concept of the 'pre-embryo' in the UK (see note 21).

certain features as belonging to the 'objective' embryo which exists as an *a priori* biological entity[f] and those features which can be characterised as belonging to the embryo as a constructed concept. It will be shown that it is often the constructed embryo that gives rise to normative accounts of what it is we may do to it or, in other words, what (legal and/or moral) 'status' it has. This goes some way to explaining the various legal constructions of the embryo and the seemingly inconsistent accounts of what we may do to 'the embryo' in different contexts. It will also be suggested that in a philosophical setting, there is a correlation between the creation of the concept and the ethical and moral position assumed via this construction. It is largely through construction that we see the embryo as an entity with the possibilities of identity, and the question of how it will be constituted by alternative future possibilities. When analysing how we can or should regulate for the embryo, it is important to base our considerations on a rich depiction of the embryo. Shapiro provides the following justification for the importance of this sort of exploration:

> Our moral pluralism requires us to understand, in a demanding but thoughtful and empathetic manner, the serious concerns of those who disagree with us. This is especially important when issues arise in an area which is already politically polarized. Indeed, no area in American political life is quite as rigidly polarized as the moral status of the embryo.[29]

3. HOW IS THE EMBRYO CONSTRUCTED?

The theme of what an embryo is, unsurprisingly, dominates much of the debate on biotechnologies. The embryo is described in many and varied

---

[e] Implicit in my discussion is the idea of the embryo as an entity. This individualist focus could in itself be contested. See note 25 (especially Chapter 6). Thanks to Marie Fox and Marie Jacob for bringing this work to my attention.

[f] It is noted that the debate over a fixed biological definition of the embryo does not negate the notion of the embryo as a biological entity for the purposes of my work (see notes 26–28).

ways. This was highlighted in the Irish case of *M.R v T.R* by the accounts given in evidence as to what the embryo was, per McGovern J:

> In opening the case Mr. Hogan for the plaintiff stated that the court "... in the context of this case, is confronted with the most difficult decision of all, a decision which probably or at least to date, eluded the most gifted and brilliant of medical scientists, ... and those in the medical community and indeed in other communities who have reflected on this; at what point does human life begin?" It is possible for Scientists and Embryologists to describe in detail the process of development from the ovum to the embryo and on to the stage when it becomes a foetus after implantation of the embryo in the wall of the uterus, but in my opinion, it is not possible for this Court to state when human life begins. The point at which people use the term 'human being' or ascribe human characteristics to such genetic material depends on issues other than science and medicine. For example, it is a matter which may be determined by one's religious or moral beliefs and, even within different religions, there can be disagreements as to when genetic material becomes a 'human being'. But it is not the function of the courts to choose between competing religious and moral beliefs.[30]

This is a common theme in discussions of the embryo. This difficulty can be seen as arising from the linkage between how we describe the embryo and whether we believe the embryo is or wish it to be seen as something that is a member of our moral community. The following are examples of the different ways in which the concept of the embryo may be constructed. The embryo as:

- a human being[31]
- a potential person[32]
- an unborn child[33]
- as a thing[34] or stuff[35]
- a 'moral work object'[36]
- a 'legally defined entity'[37]

These assorted constructions arise as a result of different background conditions against which the embryo is being discussed or considered and also as a result of interpreting particular features of the embryo in different ways. Different descriptions are acceptable to different groups for different reasons. They are controversial and subject to disagreement, and can lead to conflicting normative conclusions of what the embryo is and what it is permissible to do with it or to it. I will now analyse the descriptions of the embryo listed above in order to draw out the important aspects of each description. In doing this, the implicit ethico–legal principles contained in each description become apparent.

3.1 A Human Being[38]

There are many who believe that a human embryo is a human being equivalent in value to you or me.[39] They believe that from the moment of conception the embryo is deserving of our full respect, and it follows from this that the destruction of the embryo is wrong. Sandel summarises the views of this position as follows:

> Each one of us began life as an embryo. If our lives are worthy of respect, and hence inviolable, simply by virtue of our humanity, one would be mistaken to think that at some younger age or earlier stage of development we were not worthy of respect. Unless we can point to a definitive moment in the passage from conception to birth that marks the emergence of the human person, this argument claims, we must regard embryos as possessing the same inviolability as fully developed human beings.[g,40]

There are a few important aspects of this quote worth further attention. The first point to note is the way in which the embryo is constructed as something that is 'like us' and therefore worthy of respect.[h] This method of constructing paradigm examples of that which is worthy of respect and then applying them across cases is commonly used in discussion of moral status. Legally, bringing the embryo within the realm of the 'human' has

---

[g] It should be noted that Sandel goes on to reject this position.

[h] As opposed to non-human hybrid embryos which are less deserving of respect (see note 41).

important consequences because there is a proportionate correlation between the extent to which a being is 'human' and the level of legal protection afforded to it. Indeed being able to identify those humans who are 'like us' is integral to how we define legal persons, as Karpin states:

> What does it mean to be a human person, in a world where these transformations are possible? … The legal response to such transformative possibilities has so far been to ensure their prohibition at the earliest stage — the embryonic stage. … ensuring at least for now that such *persons* can not be born.[42]

Using language in order to make things more or less 'like us' is a rhetorical tool that has been used elsewhere.[43] In the language used in debates on embryo research, the strength of this technique is evident.[44] On the one hand, 'otherness' and ways in which an embryo is unlike us are used to show why we need not attribute full moral status to the embryo.[i,45] On the other hand, the humanness of the embryo is described in order to show how an embryo is like us, *someone* as opposed to *something*.[46] The sense in which language can be used to induce feelings of identification in people is evident in all areas of the debate. For instance, consider the terms used by Enoch Powell M.P. when he called his proposed piece of legislation 'The Unborn Children (Protection) Bill'.[47]

> The repugnance with which those of many religious persuasions or none view the actions that the Bill would forbid is of a more fundamental character. It is an instinct implanted in a human society. This is the recognition by a human society of its obligations to itself, to future generations and to human nature.[48]

What is interesting in this construction (and about the legislative reaction Karpin describes) is the presumption that embryos are entities that can be brought into existence — a point which will be returned to throughout the chapter.

---

[i] Lindsay describes similar arguments being used in the arguments relating to slavery (see note 45).

3.2 A Potential Person[49]

Potentiality, in the debates on moral status, is the claim that because human embryos have the potential to become human persons, we should respect them in the same way that we would respect human persons. This argument can be considered in a number of ways. John Harris summarises one rebuttal to this argument as follows:

> … it does not follow logically, even if we accept that we are required to treat 'x' in certain ways, and even if 'a' will inevitably become 'x,' that we must treat 'a' as if it had become 'x,' at a time or at a stage prior to its having become 'x'. This is a rather cumbersome and inelegant way of making the point that acorns are not oak trees, nor eggs omelettes.[50]

There is another way in which potentiality is important in debates about what we can do to embryos that we touched on in the last construction. Many people hold different views on what we can do to embryos in different contexts. The purpose for which an embryo is intended can affect what we believe it is acceptable to do to it. For instance, it may be considered permissible to experiment on embryos that will be destroyed after 14 days, but not on those embryos that are going to be implanted in the uterus of a woman in order to achieve a pregnancy. The following statement by the Secretary of State in the debates leading up to the Human Fertilisation and Embryology Act 1990 affirms this view:

> I think that the researchers all say that it would be irresponsible for doctors to replace in the womb fertilized human eggs that had been subjected to novel procedures or tests, for these might themselves lead to abnormality.[51]

We may have particular concerns about those embryos that will one day become people, and this may cause us to act in particular ways towards those embryos. This concern is evident, for example, in amendments to the legislation that regulates liability for actions which occur prior to birth[52] and the 'welfare of the child' provision the Human Fertilisation and Embryology Act 1990 (as amended 2008).[53]

Similarly, people may view the embryo differently when its potentiality is diminished. Consider the following excerpt:

> I think an embryo is still a baby. I still think of it as a baby right from day one. Once they come together and start cell dividing, I still think that's life. ... unused material, it's stuff that can't be put back in us . . . 'cos they don't think it's good enough.[54]

This quotation is from an interview with a woman who had been asked to donate embryos to research after going through assisted reproduction. She makes an interesting distinction between the embryo that was used for *reproductive* purposes and the one that was used for *research* purposes. Her words highlight how functionality can play an important role in the levels of respect that we believe an embryo deserves, and this functional distinction is now enshrined in the Human Fertilisation and Embryology Act 1990 (as amended 2008). This functional or purposive approach to construction of the embryo will be returned to in several places throughout the chapter.

3.3 An Unborn Child[55]

Reproductive embryos can be understood legally as belonging to a much broader category — that of 'the unborn'. Mulkay describes a seven-page feature which was published by the *Daily Mirror* entitled 'Baby Special'.[56] He explores the contradictory relationship between the embryo, the unborn, and the born baby. He says:

> Whereas the case against embryo research emphasized the harm it did to minute, invisible 'unborn children', the case for such research focused on the production of real, full-scale, visible children. By the time of the final parliamentary debates, unborn children had been displaced from the media by a throng of happy, smiling babies.[57]

This extract describes how different constructions can arise out of contradictory interpretations of the *same thing*. In this way, the same *thing* can be used effectively by those on alternate sides of the debate in order to achieve opposing ends. The depiction of the embryo as an unborn child has an obviously evocative meaning: In destroying embryos, we are destroying tiny babies. The inaccuracy of this depiction is glossed over through effective imagery.[58] This imagery is also exploited by those on the pro-research side of the debate — images of future happy babies affirm the importance of research that can improve IVF and pre-implantation genetic diagnosis (PGD) technologies.

Another way in which the embryo is constructed as unborn is through what Haimes *et al.* describe as 'Baby Talk':

> An example of the ways in which 'baby talk' was the starting point for interviewees' deliberations comes from one couple who nonetheless disagreed in their conclusions.
> The woman said:
> I think an embryo is still a baby. I still think of it as a baby right from day one. Once they come together and start cell dividing, I still think that's life.
> Her husband's view was that:
> anything up to 12 weeks is really just a mixture of cells, before it starts forming into something different.
> However, for this woman, the embryo they donated for research was:
> unused material, it's stuff that can't be put back in us . . . 'cos they don't think it's good enough.
> Although 'baby talk' was the starting point for their discussion, the embryo was not seen as a baby in absolute terms, even by someone who uses this as her initial reference point: a 'not good enough' embryo is 'stuff'.[59]

For those who are going through the processes of assisted reproduction, embryos symbolise potential children.[60] The intention that they have towards the embryo may invest it with special meaning to them. This construction is also closely connected to the above discussion of the importance of the connection between the embryo and future possible children. For example, PGD as an activity is inextricably linked with concern for the future child, which is facilitated through specific guidance by the Human Fertilisation and Embryology Authority (HFEA).[61] Through the processes of PGD, we are making choices about what future children will exist and what traits these children have. In this way it is different from IVF; traditionally IVF is concerned with assisting infertile individuals to have 'a' child. PGD often helps the often already fertile to *choose* the traits of their future children.[j] The primary goal of PGD is maximising the chance of making 'healthy babies',[63] and indeed the HFEA interprets its role as one of preventing certain kinds of births.[64]

3.4 As a Thing[65] or Stuff[k,66]

A theme that has thus emerged in analysing the various constructions of the embryo is the importance of our intention. Mulkay summarises this as follows:

> [D]octors, scientists and legislators define these organisms for technical or for legal purposes, it is difficult to believe that potential parents do not often speak of them, and think of them, as unborn children.[67]

Our intentions towards the embryo can affect our construction of it, and even those parents, described in (3.2) and (3.3), who once thought of an embryo as 'theirs', use the linguistic technique of otherness to justify decisions to donate spare or unsuitable embryos for research. As Thompson states, "[m]aking an embryo into waste is an outcome and not a by-product".[68] The decision to 'discard' or 'donate' an embryo follows a process of reimagining the sort of thing it is — a transformation that is not easy.[69] Different features of *the* embryo are highlighted as being important; the embryo is no longer a potential person, it could be a source of valuable knowledge, but it is no longer 'good enough' to be 'our baby'.[l]

This account of the embryo contrasts with the earlier description of the embryo as a mini-person. The language of otherness is used here to show

---

[j] For a discussion of whether PGD is really a choice, see note 62.

[k] In this study, they identify five legal categories of embryos.

[l] For a discussion of the contested nature of classifying embryos as 'spare', see note 70.

how unlike us the embryo is. This has interesting implications for the abortion debate, as often those who describe the embryo in this way contrast the earlier embryo with the later foetus.[71] In describing an embryo as a thing or stuff, the embryo is categorised as being akin to skin cells etc. Implicit in this description is the fact that the embryo is not valuable in the way that persons are, although it might have some other value, for example, as a source of scientific knowledge.

3.5 'Boundary Object'[72] or a 'Moral Work Object'[73]

Those working in the area of embryology, embryo research, and assisted reproduction often construct the embryo in different ways. In this section, I examine two important aspects of these constructions. The first is the embryo as a 'boundary object' bringing together the worlds of embryo research and assisted reproduction. Then I will describe the embryo as a 'moral work object'.

Williams *et al.* define boundary objects as follows:

> Boundary objects inhabit several intersecting worlds ... Boundary objects are objects that are both plastic enough to adapt to local needs . . . yet robust enough to maintain a common identity across sites . . . They have different meanings in different social worlds but their structure is common enough to more than one world to make them recognizable, a means of translation.[74]

They argue that the embryo becomes a boundary object that mediates the worlds of assisted reproduction and embryo research. They describe how the embryo takes on two differing meanings in the worlds of PGD and embryonic stem cell (ESC) research. In this way, the embryo acts as a means of anchoring the difference between the two areas of practice.

> By conceptualizing embryos as boundary objects we begin to grasp how they are de-contextualized and re-contextualized within and between the 'two cultures' of ESC and PGD labs. Embryos are sometimes different things to scientists in ESC and PGD labs as: "The knowledge, skills and

> expertise of the respective groups are different and are brought to bear on different objects".[75]

However, this anchorage is not absolute, and as these two worlds converge, the embryo acts as a translational object that can bind the two activities. Both groups see the embryo as a scarce and valuable entity. In order to avoid wasting embryos, discarded embryos from PGD can become a source of embryos for ESC research. This interplay is interesting as the value of embryos that are unsuitable for implantation in PGD is realised through their use in ESC research.[76] In this way, ESC research becomes a form of 'rescue' for embryos that would otherwise be discarded. It could be argued that we respect the embryo by using it in ESC research and in using 'discarded' embryos for the purposes of ESC research, we are treating the embryo with 'special respect' as the law requires.[m]

Another way in which the embryo is variously constructed in the work environment is as a 'moral work object'. A work object is "any material entity around which people make meaning and organize their work practices".[78] Ehrich *et al.* describe what it means for an embryo to be a moral work object as follows:

> We argue that staff working in this [Assisted Conception Unit] remake the ethical and regulated social order of work with embryos locally by employing normative, accommodating, and prioritising strategies to achieve shared work goals despite diverse constructions of the embryo and micro-level work goals. Thus the embryo is constructed as a moral work object in diverse forms that allow the staff to do their work.[79]

In their study, Ehrich *et al.* describe the varied constructions of the embryo of those working in the fields of PGD and ESC research. Given the contested nature of embryos and the various values that can attach to them, those who work with embryos employ different strategies for

[m] It is worth noting that not all scientists share this view of the interchangeability of the embryo (see note 77).

realising work goals. One strategy is to distinguish between their personal views and their professional (public) views of the embryo. This distinction allows them to act in ways towards 'work' embryos that may not reflect their own personal views of 'the' embryo. By focusing attention on overall institutional values they act in ways that facilitate the overall institutional aims of their profession.[n] Professionals often focus on work goals rather than on personal questions about fundamental values.

This construct, I suggest, has an analogue at the policy level. Here we see the embryo being constructed as a politically contentious object. The embryo is seen as being the root of disagreement between many different groupings. At a policy level, this has been acknowledged since the Warnock Committee. Regulation of what we can do to the embryo must mediate between those who believe the embryo is as valuable as you or me, and those who believe it is a 'mere' cluster of cells. In order to accommodate such a diverse range of views, there is a move away from personalised judgements of the value of an embryo to the more public judgement of how best we accommodate different views. This 'public judgement' attempts to mediate a path through the diverse views and somehow facilitate the continuation of research and assisted reproductive technologies while acknowledging the very strong views different groups may have toward the embryo.

3.6 A 'Legally Defined Entity'[81]

For many, the embryo is whatever the law of the land defines it to be. A word of caution is necessary when considering the embryo as an entity that can be subject to legal reclassification. Guenin puts the case thus:

> The antennae of those whom a defender of regenerative medicine most needs to reach have been alert to detect any move to reclassify objects of moral concern en route to their sacrifice.[82]

[n] Evidence of his type of reasoning can be seen in other areas of morally contentious practice (see note 80).

Guenin continues on to describe the derision that the terms 'preimplantation embryo' and 'pre-embryo' received from those who objected to embryo experimentation and destruction in the US. These opponents viewed the term 'pre-embryo' as little more than relabeling in an attempt to hide the controversy regarding embryo destruction, overcoming this controversy by stealth.[o] In the next section, I will consider some of the constructions of the embryo that are evident in law.

4. CONSTRUCTIONS OF THE EMBRYO IN LAW

In comparison to other countries, the UK adopts a permissive approach to the regulation of assisted reproduction and embryo research. Current regulations are not underpinned by any particular moral principle, although several are important.[p] Since the Warnock Report, the notion of 'respect' for the human embryo has been considered a central aspect of the regulations, although there are no explicit guidelines as to what 'respect' means. Many, including Warnock, question how many activities that are currently allowed can be understood as respecting the embryo.[84] The lack of a unitary moral underpinning of the regulation of embryology and assisted reproduction has led many to criticise existing regulations. Yet it is also true that there has not yet been a better mode of regulation advanced than the current framework and, indeed, UK regulations are accepted internationally as being an exemplar of good regulation in the area.[85] Many of the 'morally arbitrary' cut-off points that exist in the law relating to the unborn are ethically defensible as an aspect of regulation in a morally controversial area. If we really wish for regulation that is underpinned by a singular ethical position or principle, then we would have to change the process of policy making. The passage from Bill to Act involves gaining majority support in the House of Commons and the

---

[o] Similar arguments are evident in critiques of the use of the term 'admixed' rather than 'hybrid' embryos as discussed above.

[p] The House of Commons Science and Technology Committee 'Inquiry into Human Reproductive Technologies and the Law' (Eighth Special Report of Session 2004–05) amended the House of Commons Science and Technology Committee 'Human Reproductive Technologies and the Law' (5th Report of Session 2004–5). The amended text leaves out reference to the principled underpinning to law in this area. (See note 83.)

House of Lords, both of which are made up of individuals with differing moral attitudes and beliefs. Specifically in the context of regulating human reproduction, embryology, and embryo research, it should be noted that a major feature of the HFEA is public consultation on how regulation should evolve.[86] It is also important to consider that there are many issues at play in the debate over how we should regulate embryology and embryo research. A glance through the ethical literature shows that there are many and varied positions on the legitimacy of embryology and assisted reproduction. This literature contains arguments that extend far beyond questions about the status of the embryo: there are arguments about the nature of humanity[87] and the position of women in society,[88] to name but two. Franklin describes the regulatory frameworks for embryology as being sociological in nature:

> In contrast to prominent IVF consultants such as Robert Winston who argue that medical professional judgement is too often compromised by rigid HFEA requirements, as well as to critics of the HFEA who argue that it is inconsistent and haphazard, its supporters might argue that the HFEA works effectively *because it works sociologically* to find a pragmatic path to workable social consensus.[89]

The Human Fertilisation and Embryology Act 1990 contained a definite account of the embryo that meant that the ambit of the HFEA was open to litigation as new ways to create embryos emerged. In response to this concern, it was decided that the 2008 Act would provide an open-ended definition of the embryo in order to accommodate scientific advances in the area:

> We are concerned that any legal definitions of the embryo based on the way it was created or its capabilities would either be open to legal challenge or fail to withstand technological advance. The attempt to define an embryo in the HFE Act has proved counter-productive, and we recommend that any future legislation should resist the temptation to redefine it. We consider that a better approach would be to define the forms of embryo that can be implanted and under what circumstances. Using this approach, only those forms of embryo specified by the legislation, such as those created by fertilisation, could be implanted in the womb and thereby used for reproductive purposes. Other forms of embryo would be regulated insofar as they are created and used for research purposes. (Paragraph 53).[90]

In order to avoid legal ambiguity as to the remit of the HFEA, the Science and Technology Committee suggested that the embryo be defined in a negative way; there is a strict definition of the sort of embryo that can be implanted into a woman, but there is vagueness and open-endedness outside of this reproductive context. This distinction essentially allows the Act to embody what Hennette-Vauchez, speaking of French law, describes as a 'socio–legal (non)construction of the embryo' — setting a strict definition on the sort of embryo that can be implanted into a woman's womb for the purposes of achieving a pregnancy and leaving the definition of embryo generally open-ended so that the law can accommodate advances in technology.[91] This also further facilitates what Hennette-Vauchez has described as a 'conceptual severance' of the activities of reproduction and research.

In the context of reproduction, embryos form part of the broader category of the unborn. In a research context, they form part of the broader category of human material, although they are regulated by the Human Fertilisation and Embryology Act rather than by the Human Tissue Act 2004.[92] The law in distinguishing *in vitro* embryos must walk a difficult line between the activities of reproduction and research.[93–96] In order to walk this line, the Human Fertilisation and Embryology Act 1990 (as amended 2008) categorises embryos as the 'permitted' reproductive embryo and the 'unpermitted' (or in the words of Karpin 'prohibited') research embryo:

> (2) No person shall place in a woman —
>   (a) an embryo other than a permitted embryo (as defined by section 3ZA)[97]
>
> (4) An embryo is a permitted embryo if —
> - it has been created by the fertilisation of a permitted egg by permitted sperm,
> - no nuclear or mitochondrial DNA of any cell of the embryo has been altered, and
> - no cell has been added to it other than by division of the embryo's own cells.[98]

Defining embryos in accordance with their capacity to one day become 'beings' is an approach that has recently been adopted by the European

Court of Justice in the case of *Oliver Brüstle v Greenpeace*.[99] In this case, the German Bundesgerichtsh sought clarification from the European Court of Justice as to the meaning of the word 'embryo' for the purposes of European Union (EU) Directives which related to patenting. This followed a challenge to patents held by Oliver Brüstle (OB) in Germany. OB had patented:

- 'isolated and purified neural precursor cells'
- processes for producing these cells from embryonic stem cells
- use of these cells in treatment

Greenpeace objected to the patents and brought an action for their annulment on the grounds that because of the use of embryos involved, the patents were invalid. The domestic court of first instance had found the patents to be invalid, OB appealed, and the Bundesgerichtshof sought clarification from the Court of Justice. They asked for an opinion on the various issues including what is meant by the term embryo.

The Opinion offered in the preliminary ruling by the Advocate General (AG) is interesting and significant on many levels, not least the impact on patent laws across the EU should the Court of Justice follow the AG. Here, however, I am only interested in the approach the Court took to defining what an embryo is.

Like many before him, the AG stressed that his role was as legal, not moral, arbiter of the questions before him — this echoes the approach of McGovern J quoted above. He said: "I do not intend to decide between beliefs or to impose them."[100] Despite his avowed lack of engagement with the moral questions, he continues "[t]he body exists, is formed and develops independently of the person who occupies it".[101] In this statement, the Court, whether knowingly or not, endorses the view that whatever value attaches to the later person which results from the embryo, this does not apply to the embryo itself, although the relationship between the two can mean that the embryo is worthy of some protection. And it is this 'non-moral' foundation upon which the AG bases his opinion.

In informing his opinion, the AG found not alone was there no consistency in defining the embryo across jurisdictions, discrepancies also arose within jurisdictions. Despite the lack of consensus he identified two broad camps within which definitions could be located:

> Embryo exists from fertilisation
> Embryo exists from implantation

In order to fully appreciate the opinion issued in this case, it will be useful to recap on some themes that have been identified in this chapter. The first of these was the claim that embryos are constructed and that construction involves an interpretation of different features of the 'core embryo'. In this case, the AG had high hopes for the 'objective embryo':

> In my view, against this background only legal analyses based on objective scientific information can provide a solution which is likely to be accepted by all the Member States. The same concern for objectivity leads me to say that science's silences or its failure to provide proof are also objective information which can form the basis for a legal analysis.[102]

The second theme, identified in this chapter, is the importance of context and also the purposive way in which the embryo is often constructed. The definition in this case had, for the sake of consistency and efficacy, to be applicable across EU member states, otherwise patents valid in one jurisdiction would be invalid in another. Hence the reluctance of the AG to deal with moral questions of when life begins; rather, he is concerned with when life begins in a physical sense:

> The provisions of Directive 98/44 provide an important indication. What should be defined? The appearance of life? The amazing moment when, *in utero*, what was perhaps only a group of cells changes in nature and becomes, whilst not yet a human being, an object, or even a subject of law? Not at all. This is not the question which follows from the wording and the approach taken by the directive which, through the wise wording it uses, leads us to define not life, but the human body. It is 'the human body, at the various stages of its formation and development' for which it demands protection (31) when it declares it expressly unpatentable.[103]

In proposing a definition of the embryo, the AG states that any definition of the embryo will be time limited (as outlined above in relation to the HFEA) and context dependent. The concern here is Intellectual Property, not other areas of law making, and so the AG paves the way for a multiplicity of legal definitions of the embryo at EU level:

> I think that it is also worth pointing out that the legal definition which I will propose falls within the framework of the technical directive under examination and that, in my view, legal inferences cannot also be drawn for other areas which relate to human life, but which are on an entirely different level and fall outside the scope of Union law. For that reason, I consider that the reference made at the hearing to judgments delivered by the European Court of Human Rights on the subject of abortion is, by definition, outside the scope of our subject. It is not possible to compare the question of the possible use of human embryos for industrial or commercial purposes with national laws which seek to provide solutions to individual difficult situations.[104]

So, what is the embryo according to the AG? Well, the opinion issued relies heavily on the idea of physical continuity of the entity. As long as a cell has the possibility of one day developing into a human being, then it constitutes part of the human body at a stage in development. Thus, in accordance with the prohibition on patenting the human body at any stage of development, the embryo cannot be patented. The physical relationship between the embryo and future person is key to the definition offered by the AG, and the finding that the patentability of embryos be prohibited because the embryo constitutes stages in the development of the human body. In this case it was decided that:

> … Article 6(2)(c) of Directive 98/44 must be interpreted to the effect that the concept of a human embryo applies from the fertilisation stage to the initial totipotent cells and to the entire ensuing process of the development and formation of the human body. That includes the blastocyst. I will also argue that unfertilised ova into which a cell nucleus from a mature human cell has been transplanted or whose division and further development have been stimulated by parthenogenesis are also included in the concept of a human embryo in so far as the use of such

> techniques would result in totipotent cells being obtained. On the other hand, I will show that pluripotent embryonic stem cells are not included in that concept because they do not in themselves have the capacity to develop into a human being.[105]

This is the scientifically objective account of the embryo that the AG confines his definition to. It is not my purpose to criticise the opinion of the AG but rather to cast light on the construction of the embryo involved and so evidence the implications that different constructions of the embryo have in law.

The case is an example of the reality of how the law will continue to approach questions relating to embryos. Specific features of the embryo will be important in different contexts. Different constructions of the embryo will emerge. Ultimately, speaking of *the* embryo is fruitless and so too are debates which focus on the status of *the* embryo as an answer to questions about the permissibility of different activities. How the law defines the embryo is continually open to controversy and change. Franklin, defending the approach of the Warnock Committee, summarises as follows:

> Like rights, embryos exist as a plurality. It is not possible to give them an 'absolute status' — legally or ethically any more than socially or politically.[106]

This criticism can be taken further, and we could challenge the utility of moral status as a concept in law. To rely on arguments about moral status to guide regulation of the embryo is, to quote English, "to clarify *obscurum per obscurius*".[q,107]

5. CONCLUSION

In order to reach a consensus about how to regulate for the embryo, we first need to take account of what the embryo is. As Haimes *et al.* conclude:

---

[q] See Sorcha Uí Chonnachtaigh's chapter in this volume for more on moral status.

> [I]t is not possible to have an effective debate about the acceptability of embryo research if discussion is restricted to the moral status of the abstract embryo; the real physical entities that reside in the freezer, clinic and women's bodies need to be considered.[108]

The 'construction' of the embryo can have serious consequences for how we regulate assisted reproductive technologies and embryology. Legally and morally, the origin and position of the embryo can affect our view of what it is and what we can do with it.[r] The perceived moral, and legal, status of the embryo will be, by definition, influenced by our construction of it.[112,113] Various constructions of the embryo are evident in the literature on the ethics and law of assisted reproduction and embryology. Often, these constructions reflect a distinction between different *types* of embryo — 'the research embryo' or 'the reproductive embryo' or 'hybrid embryos'. In these various constructions, it is evident that certain characteristics belong to the 'objective' embryo, for example, the embryo is 'a cell mass which is not sentient'. Other features can be characterized as belonging to the embryo as a constructed concept, for example, the embryo as 'my future child' or as 'a symbol of human life'. All these descriptions of the embryo can lead to normative conclusions of what the embryo is that are at odds with each other. There is often a conflation of descriptive account of the embryo and this account having normative value or, as Haimes *et al.* describe, there is an "inextricable entangling of the moral and social status of embryos".[114]

By understanding the embryo as an essentially constructed entity, disagreements about what we can do to the embryo and about embryonic stem cell research may become more understandable. The starting point must be to understand what it is people are disagreeing about, before moving to questions of how we deal with the different approaches to the embryo. Haimes summarises the challenges faced by the varied understanding of the embryo:

> Both studies reveal a multiplicity of definitions, identities and meanings of the embryo held by interviewees, legislators and ethicists. Across these

[r] Consider the distinction between embryos, pre-embryos, and clonotes (see notes 109–111).

> different discursive domains we can see differences between: 'the embryo' as an abstraction with an emblematic moral status; 'an embryo', which could be any particular embryo that results from the IVF process; and 'our embryo', which is the embryo produced through IVF and imbued with particular social and moral values by the couple being treated (because 'our embryo' might become 'our baby').[115]

The varied meanings of the embryo in different contexts, including for use in hESC research, require us to take a broad range of issues into account when we decide how it is we regulate for the embryo. By examining the way in which the embryo is constructed, we can come to a clearer idea of what aspects of the construction are important to different groups. That is not to say that taking a constructionist approach will overcome the differences in approach to the embryo — rather it will enable us to better understand the disagreements and move policy forward in a way that can better accommodate these differing views.

REFERENCES

1. Carroll L. (1865/1998) *Alice's Adventures in Wonderland.* Penguin Classics, London, p. 61.
2. Tooley M. (1983) *Abortion and Infanticide.* Clarendon Press, Oxford, p. 12.
3. Franklin S. (2010) Response to Marie Fox and Thérèse Murphy. *Social & Legal Studies* **19:** 505–509, p. 505.
4. *Oliver Brüstle v Greenpeace* (C-34/10) (2010/C 100/29).
5. Mulkay M. (1997) *The Embryo Research Debate.* Cambridge University Press, Cambridge.
6. Parry S. (2003) The Politics of Cloning: Mapping the Rhetorical Convergence of Embryos and Stem Cells in Parliamentary Debates. *New Genetics and Society* **22:** 177–200.
7. Williams C, *et al.* (2003) Envisaging the Embryo in Stem Cell Research: Rhetorical Strategies and Media Reporting of Ethical Debate. *Sociology of Health and Illness* **25:** 793–814.
8. Franklin S. (2006) Embryonic Economies: The Double Reproductive Value of Stem Cells. *Biosocieties* **1:** 71–90.
9. Salter B. (2007) *Bioethics,* Politics and the Moral Economy of Human Embryonic Stem Cell Science: The Case of the European Union's Sixth Framework Programme. *New Genetics and Society* **26:** 269–288.

10. Pardo R, Calvo F. (2008) Attitudes towards Embryo Research, Worldviews and the Moral Status of the Embryo Frame. *Science Communications* **30:** 8–47.
11. Williams C, Wainwright SP, Ehrich K, Michael M. (2008) Human Embryos as Boundary Objects? Some Reflections on the Biomedical Worlds of Embryonic Stem Cells and Pre-Implantation Genetic Diagnosis. *New Genetics and Society* **27:** 7–18.
12. Haimes E, Porz R, Scully J, Rehmann-Sutter C. (2008) 'So What is an Embryo?' A Comparative Study of the Views of those Asked to Donate Embryos for hESC Research in the UK and Switzerland. *New Genetics and Society* **27:** 113–126.
13. Ehrich K, Williams C, Farsides B. (2008) The Embryo as Moral Work Object: PGD/IVF Staff Views and Experiences. *Sociology of Health and Illness* **30:** 772–787.
14. Jacob M, Prainsack B. (2010) Embryonic Hopes: Controversy, Alliance, and Reproductive Entities in Law and the Social Sciences. *Social & Legal Studies* **19:** 497–517.
15. Ehrich K, Williams C, Farsides B. (2008) The Embryo as Moral Work Object: PGD/IVF Staff Views and Experiences. *Sociology of Health and Illness* **30:** 772–787, p. 780.
16. Devolder K, Harris J. (2007) The Ambiguity of the Embryo: Ethical Inconsistency in the Human Embryonic Stem Cell Debate. *Metaphilosophy* **38:** 153–169.
17. Potter J. (2003) *Representing Reality: Discourse, Rhetoric and Social Construction.* Sage, London, Chapter 4.
18. On the 'emerging pro-science apenda' see Fox M. (2008) Legislating Interspecies Embryos. In: Smith SW, Deazley R (eds.), *The Legal, Medical and Cultural Regulation of the Body.* Ashgate, Surrey, p. 105 onwards.
19. Silver L. (1999) *Remaking Eden: Cloning, Genetic Engineering and the Future of Humankind?* Phoenix Giant, London, p. 46.
20. Williams C, *et al.* (2003) Envisaging the Embryo in Stem Cell Research: Rhetorical Strategies and Media Reporting of Ethical Debate. *Sociology of Health and Illness* **25:** 793–814, p. 800.
21. Mulkay M. (1994) The Triumph of the Pre-Embryo: Interpretations of the Human Embryo in Parliamentary Debate over Embryo Research. *Social Studies of Science* **24:** 611–639.
22. Department of Health. (2007) Government Response to the Report from the House of Commons Science and Technology Committee: Government proposals for the regulation of hybrid and chimera embryos. Available at http://www.dh.gov.uk/en/Publicationsandstatistics/Publications/PublicationsPolicyAndGuidance/DH_075668 (Accessed 16 June 2011).

23. Sample I. (5 January 2007) Scientists attack plan to ban 'hybrid' embryos. *The Guardian.* Available at http://www.guardian.co.uk/science/2007/jan/05/medicalresearch.health (Accessed 16 June 2011).
24. Human Genetics Alert. 'GM and Human–Animal Hybrid Embryos — confused about the difference?' Available at http://www.hgalert.org/GM_and_hybrid_embryos_confused.html (Accessed 16 June 2011).
25. Strathern M. (1992) *Reproducing the Future: Anthropology, Kinship, and the New Reproductive Technologies.* Manchester University Press, Manchester, Chapter 6.
26. Findlay JK, *et al.* (2007) Human Embryo; A Biological Definition. *Human Reproduction* **22:** 905–911.
27. Kiesling A. (2004) What is an Embryo? *Connecticut Law Review* **36:** 1051–1092.
28. Guenin LM. (2004) On Classifying the Developing Organism. *Connecticut Law Review* **36:** 1115–1131.
29. Shapiro HT. (2004) What is an Embryo? Commentary. *Connecticut Law Review* **36:** 1093–1097, p. 1093.
30. *M R -v- T R & Ors* [2006] IEHC 359 at 363.
31. George RP, Lee P. (Fall 2004/Winter 2005) Embryos and Acorns. *The New Atlantis* **7:** 90–100.
32. Szawarski Z. (1996) Talking About Embryos. In: Evans D (ed.), *Conceiving the Embryo: Ethics, Law, and Practice in Human Embryology.* Nijhoff, The Hague, pp. 119–134, p. 119.
33. Haimes E, Porz R, Scully J, Rehmann-Sutter C. (2008) 'So What is an Embryo?' A Comparative Study of the Views of those Asked to Donate Embryos for hESC Research in the UK and Switzerland. *New Genetics and Society* **27:** 113–126.
34. Szawarski Z. (1996) Talking About Embryos. In: Evans D (ed.), *Conceiving the Embryo: Ethics, Law, and Practice in Human Embryology.* Nijhoff, The Hague, pp. 119–134, p. 119.
35. Haimes E, Porz R, Scully J, Rehmann-Sutter C. (2008) 'So What is an Embryo?' A Comparative Study of the Views of those Asked to Donate Embryos for hESC Research in the UK and Switzerland. *New Genetics and Society* **27:** 113–126.
36. Ehrich K, Williams C, Farsides B. (2008) The Embryo as Moral Work Object: PGD/IVF Staff Views and Experiences. *Sociology of Health and Illness* **30:** 772–787.
37. Human Fertilisation and Embryology Act 1990 (as amended 2008), s.1.

38. George RP, Lee P. (Fall 2004/Winter 2005) Embryos and Acorns. *The New Atlantis* **7:** 90–100.
39. George RP, Tollefsen C. (2008) *Embryo: A Defense of Human Life.* The Doubleday Broadway Publishing Group, New York.
40. Sandel M. (2004) Embryo Ethics — The Moral Logic of Stem-Cell Research. *New England Journal of Medicine* **351:** 207–209, pp. 207–208.
41. Fox M. (2008) Legislating Interspecies Embryos. In: Smith SW, Deazley R (eds.), *The Legal, Medical and Cultural Regulation of the Body.* Ashgate, Surrey, pp. 95–126.
42. Karpin I. (2006) The Uncanny Embryos: Legal Limits to the Human and Reproduction Without Women. *Sydney Law Review* **28:** 599–623, p. 611.
43. Nussbaum M. (1995) Objectification. *Philosophy & Public Affairs* **24:** 249–291.
44. Devolder K. (2006) What's in a Name? Embryos, Entities, and ANTities in the Stem Cell Debate. *Journal of Medical Ethics* **32:** 43–48.
45. Lindsay RA. (2005) Slaves, Embryos, and Nonhuman Animals: Moral Status and the Limitations of Common Morality Theory. *Kennedy Institute of Ethics Journal* **15:** 323–346.
46. Fox M. (2000) Pre-persons, Commodities or Cyborgs: The Legal Construction and Representation of the Embryo. *Health Care Analysis* **8:** 171–188.
47. The Unborn Children (Protection) Bill, 15 February 1985. Available at http://hansard.millbanksystems.com/bills/unborn-children-protection-bill (Accessed 16 June 2011).
48. Powell E. (15 February 1985) House of Commons Debate, vol 73, cc637–702, 641.
49. Szawarski Z. (1996) Talking About Embryos. In: Evans D (ed.), *Conceiving the Embryo: Ethics, Law, and Practice in Human Embryology.* Nijhoff, The Hague, pp. 119–134, p. 119.
50. Harris J. (1999) The Concept of the Person and the Value of Life. *Kennedy Institute of Ethics Journal* **9:** 293–308, p. 297.
51. Clarke K. (23 April 1990) Commons col.31, as quoted in: Mulkay M. (1997) *The Embryo Research Debate.* Cambridge University Press, Cambridge.
52. Congenital Disabilities (Civil Liability) Act 1976, s.1A.
53. Human Fertilisation and Embryology Act 1990 (as amended 2008), s.13(5).
54. Haimes E, Porz R, Scully J, Rehmann-Sutter C. (2008) 'So What is an Embryo?' A Comparative Study of the Views of those Asked to Donate Embryos for hESC Research in the UK and Switzerland. *New Genetics and Society* **27:** 113–126.

55. Haimes E, Porz R, Scully J, Rehmann-Sutter C. (2008) 'So What is an Embryo?' A Comparative Study of the Views of those Asked to Donate Embryos for hESC Research in the UK and Switzerland. *New Genetics and Society* **27:** 113–126.
56. Mulkay M. (1997) *The Embryo Research Debate.* Cambridge University Press, Cambridge, p. 74.
57. Mulkay M. (1997) *The Embryo Research Debate.* Cambridge University Press, Cambridge, p. 74.
58. Franklin S. (2006) *Born and Made: An Ethnography of Preimplantation Genetic Diagnosis.* Princeton University Press, Princeton, Chapter 1.
59. Haimes E, Porz R, Scully J, Rehmann-Sutter C. (2008) 'So What is an Embryo?' A Comparative Study of the Views of those Asked to Donate Embryos for hESC Research in the UK and Switzerland. *New Genetics and Society* **27:** 113–126, p. 116 [reference omitted].
60. Jacob M, Prainsack B. (2010) Embryonic Hopes: Controversy, Alliance, and Reproductive Entities in Law and the Social Sciences. *Social & Legal Studies* **19:** 497–517.
61. Human Fertilisation and Embryology Act 1990 (as amended 2008), s.14.
62. Franklin S. (2006) *Born and Made: An Ethnography of Preimplantation Genetic Diagnosis.* Princeton University Press, Princeton, Chapter 3.
63. Williams C, Wainwright SP, Ehrich K, Michael M. (2008) Human Embryos as Boundary Objects? Some Reflections on the Biomedical Worlds of Embryonic Stem Cells and Pre-Implantation Genetic Diagnosis. *New Genetics and Society* **27:** 7–18, p. 10.
64. See guidance on dysgenic selection Human Fertilisation and Embryology Authority, Code of Practice (8th Edition), section10c.
65. Szawarski Z. (1996) Talking About Embryos. In: Evans D (ed.), *Conceiving the Embryo: Ethics, Law, and Practice in Human Embryology.* Nijhoff, The Hague, pp. 119–134, p. 119.
66. Haimes E, Porz R, Scully J, Rehmann-Sutter C. (2008) 'So What is an Embryo?' A Comparative Study of the Views of those Asked to Donate Embryos for hESC Research in the UK and Switzerland. *New Genetics and Society* **27:** 113–126.
67. Mulkay M. (1994) The Triumph of the Pre-Embryo: Interpretations of the Human Embryo in Parliamentary Debate over Embryo Research. *Social Studies of Science* **24**: 611–639, p. 634.
68. Thompson C. (2005) *Making Parents: The Ontological Choreography of Reproductive Technologies.* The MIT Press, Massachusetts, p. 264.

69. Haimes E, Porz R, Scully J, Rehmann-Sutter C. (2008) 'So What is an Embryo?' A Comparative Study of the Views of those Asked to Donate Embryos for hESC Research in the UK and Switzerland. *New Genetics and Society* **27:** 113–126, p. 125.
70. Ehrich K, Williams C, Farsides B. (2010) Fresh or Frozen? Classifying 'Spare' Embryos for Donation to Human Embryonic Stem Cell Research. *Social Science and Medicine* **71:** 2204–2211.
71. Parry S. (2003) The Politics of Cloning: Mapping the Rhetorical Convergence of Embryos and Stem Cells in Parliamentary Debates. *New Genetics and Society* **22:** 177–200.
72. Williams C, Wainwright SP, Ehrich K, Michael M. (2008) Human Embryos as Boundary Objects? Some Reflections on the Biomedical Worlds of Embryonic Stem Cells and Pre-Implantation Genetic Diagnosis. *New Genetics and Society* **27:** 7–18, p. 10.
73. Ehrich K, Williams C, Farsides B. (2008) The Embryo as Moral Work Object: PGD/IVF Staff Views and Experiences. *Sociology of Health and Illness* **30:** 772–787.
74. Williams C, Wainwright SP, Ehrich K, Michael M. (2008) Human Embryos as Boundary Objects? Some Reflections on the Biomedical Worlds of Embryonic Stem Cells and Pre-Implantation Genetic Diagnosis. *New Genetics and Society* **27:** 7–18, p. 8 [Quoting Star and Griesemer].
75. Williams C, Wainwright SP, Ehrich K, Michael M. (2008) Human Embryos as Boundary Objects? Some Reflections on the Biomedical Worlds of Embryonic Stem Cells and Pre-Implantation Genetic Diagnosis. *New Genetics and Society* **27**: 7–18, p. 15.
76. Woods S. (2008) Stem Cell Stories: From Bedside to Bench. *Journal of Medical Ethics* **34:** 845–848.
77. Franklin S. (2006) Embryo Economies: The Double Reproductive Value of Stem Cells. *Biosocieties* **1:** 71–90.
78. Casper MJ. (1998) *The Making of the Unborn Patient: A Social Anatomy of Fetal Surgery.* Rutgers University Press, London. As quoted in Ehrich K, Williams C, Farsides B. (2008) The Embryo as Moral Work Object: PGD/IVF Staff Views and Experiences. *Sociology of Health and Illness* **30:** 772–787, p. 775.
79. Ehrich K, Williams C, Farsides B. (2008) The Embryo as Moral Work Object: PGD/IVF Staff Views and Experiences. *Sociology of Health and Illness* **30:** 772–787, p. 783.
80. Farsides B, Williams C, Alderson P. (2004) Aiming Towards 'Moral Equilibrium': Health Care Professionals' Views of Working Within the Morally Contested Field of Antenatal Screening. *Journal of Medical Ethics* **30:** 505–509.

81. Human Fertilisation and Embryology Act 1990 (as amended 2008), s.1.
82. Guenin LM. (2004) On Classifying the Developing Organism. *Connecticut Law Review* **36:** 1115–1131, p. 1124.
83. House of Commons Science and Technology Committee 'Inquiry into Human Reproductive Technologies and the Law' (Eighth Special Report of Session 2004–05) Available at http://www.publications.parliament.uk/pa/cm200405/cmselect/cmsctech/491/491.pdf (Accessed 16 June 2011).
84. As quoted in Jackson E. (2006) Fraudulent Stem Cell Research and Respect for the Embryo. *BioSocieties* **1:** 349–356.
85. Scott R. (2007) *Choosing Between Possible Lives: Law and Ethics of Prenatal and Preimplantation Genetic Diagnosis.* Hart Publishing, Oxford & Portland, OR, p. 256.
86. Montgomery J. (1991) Rights, Restraints and Pragmatism: The Human Fertilisation and Embryology Act 1990. *Modern Law Review* **54:** 524–534.
87. Sandel M. (2004) Embryo Ethics — The Moral Logic of Stem-Cell Research. *New England Journal of Medicine* **351:** 207–209, pp. 207–208.
88. Barnett I, Steuernagel T. (7 April 2005) Framing Assisted Reproduction: Feminist Voices and Policy Outcomes in Germany and the U.S. Paper presented at the annual meeting of The Midwest Political Science Association. Palmer House Hilton, Chicago, Illinois.
89. Franklin S. (2006) *Born and Made: An Ethnography of Preimplantation Genetic Diagnosis.* Princeton University Press, Princeton, pp. 19–20.
90. HM Government. (2005) Government Response to the Report from the House of Commons Science and Technology Committee: Human Reproductive Technologies and the Law (The Stationary Office Norwich). Available at http://www.dh.gov.uk/prod_consum_dh/groups/dh_digitalassets/@dh/@en/documents/digitalasset/dh_4117874.pdf (Accessed 16 June 2011).
91. Hennette-Vauchez S. (2009) Words Count: How Interest in Stem Cells Has Made the Embryo Available — A Look at the French Law on Bioethics. *Medical Law Review* **17:** 52–75, p. 54.
92. Human Tissue Act 2004, s.53.
93. Brazier M. (1998) Embryos' 'Rights': Abortion and Research. In: Freeman MDA (ed.), *Medicine, Ethics and the Law.* Stevens & Sons, London, pp. 9–22.
94. Fox M. (2000) Pre-persons, Commodities or Cyborgs: The Legal Construction and Representation of the Embryo. *Health Care Analysis* **8:** 171–188.
95. Ford M. (2008) Law's Ambivalent Response to Transformation and Transgression at the Beginning of Life. In: Smith SW, Deazley R (eds.), *The Legal, Medical and Cultural Regulation of the Body.* Ashgate, Surrey, pp. 21–46.

96. Fox M. (2008) Legislating Interspecies Embryos'. In: Smith SW, Deazley R (eds.), *The Legal, Medical and Cultural Regulation of the Body*. Ashgate, Surrey, pp. 95–126.
97. Human Fertilisation and Embryology Act 1990 (as amended 2008), s3(2).
98. Human Fertilisation and Embryology Act 1990 (as amended 2008), Schedule 3ZA.
99. *Oliver Brüstle v Greenpeace* (C-34/10) (2010/C 100/29).
100. *Oliver Brüstle v Greenpeace* (C-34/10) (2010/C 100/29), p. 40.
101. *Oliver Brüstle v Greenpeace* (C-34/10) (2010/C 100/29), p. 73.
102. *Oliver Brüstle v Greenpeace* (C-34/10) (2010/C 100/29), p. 47.
103. *Oliver Brüstle v Greenpeace* (C-34/10) (2010/C 100/29), p. 72.
104. *Oliver Brüstle v Greenpeace* (C-34/10) (2010/C 100/29), p. 49.
105. *Oliver Brüstle v Greenpeace* (Case C-34/10) (2010/C 100/29).
106. Franklin S. (2010) Response to Marie Fox and Thérèse Murphy. *Social & Legal Studies* **19(4):** 505–510, p. 508.
107. English J. (1975) Abortion and the Concept of a Person. *Canadian Journal of Philosophy* **5:** 236.
108. Haimes E, Porz R, Scully J, Rehmann-Sutter C. (2008) 'So What is an Embryo?' A Comparative Study of the Views of those Asked to Donate Embryos for hESC Research in the UK and Switzerland. *New Genetics and Society* **27:** 113–126, p. 125.
109. Cohen CB, Brandhorst BP. (2 January 2008) Getting Clear on the Ethics of iPS Cells. *Bioethics Forum Blog*. Available at http://www.thehastingscenter.org/Bioethicsforum/Post.aspx?id=710 (Accessed 16 June 2011).
110. George RP, Lee P. (Fall 2004/Winter 2005) Embryos and Acorns. *The New Atlantis* **7:** 90–100.
111. Agar N. (2007) Embryonic Potential and Stem Cell. *Bioethics* **21:** 198–207.
112. Fox M (2000) Pre-persons, Commodities or Cyborgs: The Legal Construction and Representation of the Embryo. *Health Care Analysis* **8:** 171–188.
113. Bortolotti L, Harris J. (2006) Embryos and Eagles: Symbolic Value in Research and Reproduction. *Cambridge Quarterly of Healthcare Ethics* **15:** 22–34.
114. Haimes E, Porz R, Scully J, Rehmann-Sutter C. (2008) 'So What is an Embryo?' A Comparative Study of the Views of those Asked to Donate Embryos for hESC Research in the UK and Switzerland. *New Genetics and Society* **27:** 113–126, p. 125.
115. Haimes E, Porz R, Scully J, Rehmann-Sutter C. (2008) 'So What is an Embryo?' A Comparative Study of the Views of those Asked to Donate Embryos for hESC Research in the UK and Switzerland. *New Genetics and Society* **27:** 113–126, p. 124.

4

Legal Regulation of Human Stem Cell Technology

Loane Skene

Stem cell technologies are subject to varying degrees of legal regulation around the world. This chapter examines regulation and policy in Australia and the United Kingdom and looks at how legislation deals with constantly changing scientific advances in the stem cell arena. It outlines the origins of the legislation and then notes some recent developments in stem cell science and treatment. Following this, the chapter considers the regulatory landscape in light of these developments, including the recent Heerey review in Australia, paying particular attention to the impact of induced pluripotent stem cells and to potential future developments in stem cell science. Overall, it is argued that the current law operates well and that minimal legislative change is required.

1. INTRODUCTION

Human stem cell technology holds great promise for human health care. In pure science, it is already revealing valuable information about the operation and function of cells in the body, the development of early human embryos and possible abnormalities that may cause or contribute to birth defects. Scientists can check the effectiveness of new drugs by

Melbourne Law School, University of Melbourne, Victoria 3010, Australia. Email: l.skene@unimelb.edu.au extracting and multiplying cells from patients with particular diseases, creating a 'disease in a dish' to study. In future, it is hoped that it will be possible to treat patients with conditions like heart disease, diabetes, spinal and brain injuries, and Alzheimer's disease by transplanting stem cells derived from their own bodies to stimulate the repair of diseased tissue. Research using stem cell treatment with animals has had some encouraging results, and new stem cell treatments have recently been started for human patients, using stem cells derived from a patient's own body cells. The first tissue-matched body parts have been developed (in recent reported cases of new tracheas) by transplanting a person's stem cells onto an artificial 'scaffold'. In future, this technique may be used to replace diseased organs in the patient's body.

The concept of treating patients by transplanting stem cells is not new. Indeed it has been an established treatment for more than 30 years in treating patients with blood disorders like leukaemia. Bone marrow, a type of body tissue containing stem cells, is obtained from donors and transplanted into patients. But, if the transplanted cells can be obtained from the patient rather than a donor, the cells are less likely to be rejected as foreign material by the patient's immune system, and the patient may avoid a life-time of immunosuppressive drugs which often have adverse effects on the patient's body and quality of life.

Many people believe that the new human stem cell technology is not so different from the established bone marrow transplants for blood disorders as to require new laws. The procedures used for those treatments are regulated principally by the relatively light hand of the common law, together with incidental legislation like that relating to quality assurance that applies to all new drugs and medical techniques, and ethical guidelines that cover medical research and experimental procedures. Yet, throughout the world, there has been much more stringent legal regulation of stem cell technology than most other novel procedures. In the United Kingdom (UK) and Australia, detailed legislation has been developed after extensive and often heated community consultation. The issues are sensitive, especially regarding the use of human embryos in research. Views are diverse, often firmly held and difficult to reconcile. There have been widespread calls for legislation with criminal sanctions for non-compliance (unusual for medical research). The statutory requirements are sometimes complex and difficult to interpret. Also, because of the speed and unpredictability of new scientific developments, the legislation needs to be constantly amended. This is time-consuming and costly. It also increases the risk of gaps and inconsistencies.

This has been apparent in Australia, where federal and state legislation was enacted in 2002: the Federal Prohibition of Cloning Act 2002 and the Research Involving Human Embryos Act 2002, with mirror Acts in the states. The federal legislation stated that it had to be reviewed three years after it came into effect and, in 2005, an extensive review was undertaken by a federal committee whose categories of members and terms of reference were stated in the two original Acts. This committee was known as the Lockhart Committee, named after its Chair, the late John Lockhart AO. The present author was Deputy Chair. The Committee produced a substantial report[1] after wide-ranging community consultation throughout the country, and the federal and state legislation was amended to implement the Committee's recommendations.[2,3] The amending legislation required that another review be undertaken three years after it became effective, which meant by the end of 2010. Thus, in December 2010, an independent legislation review committee was set up to review the Prohibition of Human Cloning for Reproduction Act 2002 and the Research Involving Human Embryos Act 2002.[4] This Committee was called the Heerey Committee, named after its Chair, the Honourable Peter Heerey QC, and the author was a member of Heerey Committee.

This chapter describes recent scientific achievements in stem cell research and then outlines some of the regulatory provisions, especially in the UK and Australia. It also discusses some of the issues that were considered by the Heerey Committee in its review of the Australian legislation and notes some of the Committee's recommendations. It is suggested that, as a general principle, there are strong arguments for not restricting scientific research by law. If scientists are left to pursue their research freely, the research that is undertaken will find 'level water'. If a particular line of research is not productive (as many people believe is currently the case with some forms of human embryonic stem cell research), the scientists will abandon those projects and try something else (for example, research on induced pluripotent stem cells or 'adult' stem cells — transplanting stem cells from one part of the body to another). If the new project yields promising results, then who would want to stop it? In other words, the restrictions on research will come from the research itself, and not from externally imposed limits such as the restrictive legislation that has been common in this area. It is acknowledged, however, that some initial restrictions may be necessary to set the broad limits by prohibiting activities that infringe on widely held values. These might include the use of reproductive cloning to breed genetically identical people, attempting to breed human–animal hybrids, and developing a human embryo outside a woman's body for more than 14 days.

2. THE EARLY LEGISLATION ON HUMAN STEM CELL RESEARCH

The main reason for enacting specific legislation in relation to human stem cell technology was that human embryos were used to derive human embryonic stem cells for the early research. Whether those embryos were 'excess' embryos formed for fertility treatment but no longer needed, or embryos created specifically for research, they were widely regarded as being morally significant (even if some people did not view them as persons, or potential persons).[a,5] Many people believed it was morally wrong to use human embryos in research, especially if they were deliberately created for research rather than being created for fertility treatment and no longer needed for that purpose. Even those who supported the research accepted the need for legislation to set the limits for research. Legislation can be enabling as well as prohibitive; scientists know that they can lawfully undertake particular activities as long as they obey 'the rules'. However, some activities were considered so reprehensible or risky that they were made criminal offences with severe penalties, like reproductive cloning and creating human–animal hybrids.

[a] For example, in Jewish ethics, an embryo is apparently regarded as an embryo only when implanted in a woman's uterus; until implanted, it is regarded as a 'pre-embryo', which may be used in research (see note 5).

The early legislation on human stem cell research tried to take account of community concerns about research on human embryos while at the same time allowing scientists to do the research. For example, in Australia, research involving human embryos is permitted if the researcher obtains a licence from the federal licensing committee and reports regularly to the licensing committee and to Parliament. The research must also be approved and monitored by an institutional Human Research Ethics Committee. Scientists must justify the use of human embryos and use as few as possible to achieve the aims of their research.

Since the early legislation was passed, it has become possible to derive tissue-matched pluripotent stem cells that could be used in human healthcare without using and destroying human embryos. These cells are called induced pluripotent stem cells (iPSCs). Scientists are also 'reprogramming' people's body cells directly into other types of cells, which might also be used to treat disease in that person without being rejected by the person's immunological system.

3. RECENT DEVELOPMENTS IN STEM CELL SCIENCE AND TREATMENT

It may be helpful at this point to examine in more detail some of the recent developments in stem cell science and treatments, as there has been significant progress.

3.1 Animals

For research involving animals, a few examples will illustrate what has been achieved to date. Note that the transplanted cells have come from embryos, from iPSCs, and from the animal's own adult body.

- Mice with a condition like Parkinson's disease in humans improved after being treated with cloned *embryonic stem cells.* Scientists inserted DNA from the mice into an enucleated egg (somatic cell nuclear transfer, or the 'Dolly technique') to form an embryo, and cultured the embryonic stem cells in the laboratory to grow into nerve cells that produce dopamine. They injected those cells into the mice's brains so that each received neurons grown from their own cloned stem cells. Their symptoms improved, and they showed no signs of rejecting the transplanted material.[6] A similar experiment with monkeys successfully reversed Parkinson-like symptoms without rejection.[7]

- Mice with a condition like sickle cell anemia in humans were treated with cells taken from their tails, reprogrammed into *iPSCs* and then differentiated into cells that produced healthy red blood cells.[8]
- Mice with a condition like muscular dystrophy in humans improved after being injected with particular adult mouse muscle stem cells, which established themselves and continued to repair and replace the damaged muscle.[9,10] In other research, a rat that could not move its left paw was able to move its paw again after a graft of stem cells from its nose, which triggered the growth of severed nerve fibres[11,12]; and an eight-year-old German Shepherd dog with osteoarthritis and hip dysplasia improved after stem cells from its abdomen were injected into its joints.[13]
- Mice have been born from mouse '*in vitro*'-derived (IVD) gametes, i.e., gametes derived from other types of cells and precursor cells from fetal tissue.[14]

3.2 Humans

In human patients, there have also been reports of some treatments that have been successful or have been claimed to be successful.[b,15–17] For example, a paralyzed man with a broken spinal cord was reported to be walking again after his stem cells (derived from his own bone marrow) were injected into the site of paralysis.[18,19] However, clinical trials are just

[b] In contrast, a press report on stem cell transplants in Costa Rica (using stem cells from patients' own bodies) stated that there is no evidence that they have been effective, and the clinic is being closed (see notes 15–17).

starting,[c,d,20,21] with limited success reported to date. Examples of treatments in human patients include the following:

- Seventeen patients (of a test group of 21) suffering from early multiple sclerosis reportedly showed "significant improvements in their condition" after being injected with stem cells from their own bone marrow by doctors in Chicago; and a "control trial . . . has been approved with 110 patients and research teams in the United States (US), Canada and Brazil".[22,23]
- In Sweden, scientists at the Karolinska Institutet have shown how "transplanted stem cells can connect with and rescue threatened neurons and brain tissue", which suggests that "a possible strategy for treating neurodegenerative diseases is to transplant stem cells into the brain that prevent existing nerve cells from dying".[24]
- In the UK, heart attack victims reportedly had "positive changes" after stem cells from their own bone marrow were injected into their damaged hearts within six hours of the attacks.[25]
- In the UK, eight patients with seriously impaired vision in one eye had their vision improved and eye pain reduced after stem cells from their 'good eye' were transplanted into the impaired eye.[26] This provides hope for similar stem cell transplants for other patients with 'bilateral damage'.[e,27–29] Patients with hearing loss have also been treated by stem cell treatment "to hair cells and neurons, deep inside the ear, that causes almost 90 per cent of hearing loss, by growing new cells and nerves".[30]
- In 2008, Spanish doctors used stem cells from bone marrow to create a whole new human organ — a trachea — for transplantation.[31,32] Also, a British boy had stem cell treatment to grow new cheekbones.[33]

---

[c] E.g., in the US, the "first human clinical trial of a stem cell treatment in ALS patients" started this year; stem cells from eight-week-old foetal tissue were injected into the spine of a man in his early 60s who has advanced amyotrophic lateral sclerosis (ALS) (see note 20).

[d] In the UK, trials of a stem cell therapy using "brain-derived stem cells… originally isolated from a human fetus" have obtained approved for stroke victims (see note 21).

[e] In the US, a company has obtained FDA approval for a trial with 12 patients and three clinical sites to treat juvenile blindness in patients with a genetic eye disease, Stargardt's Macular Dystrophy; human embryonic stem cells will be used in this treatment (see note 29).

Stem cell research is an area of extensive investment,[f,34,35] especially in the US, and it is believed to have greater potential in future. Two clinical trials involving human embryonic stem cells have been approved by the Food and Drug Administration (FDA),[g,36] and a second application to use human embryonic stem cells is pending.[37]

4. REVIEWING THE REGULATORY STRUCTURE IN LIGHT OF SCIENTIFIC PROGRESS

The developments in iPSCs and the reprogramming of body cells have been welcomed as the first real success in human stem cell research. Indeed, the magazine *Science* judged the reprogramming of adult cells as the greatest scientific breakthrough of 2008, from any area of science, saying that "[I]t actually works. It is not all spin and vague promises".[38] The new developments were also widely regarded as resolving many of the ethical issues relating to human embryo research.

However, this is open to question. Recent research on mice in the US has shown that there are "key genetic differences" between embryonic stem cells and iPSCs "which might help to explain the apparent lower robustness and differentiation efficiency of iPS [cells]".[h,39,40] There may also be different regulatory implications: "[U]nder current European and American law, each fresh set of patient-specific cells [produced by iPSC techniques] would need to be independently approved. Embryonic stem

---

[f] E.g., See CIRM press release (note 34). The press release also mentioned "Proposition 71, the California Stem Cell Research and Cures Act…which provided $3 billion in funding for stem cell research at California universities and research institutions".

[g] The first was the Geron trial using human embryonic stem cells to treat spinal cord injuries in up to 10 patients (see note 36); the second was the one for Stargardt's Macular Dystrophy (see note 37).

[h] In contrast to the report cited in note 39, another report (see note 40) put forth a more positive view: "[I]f a similar gene signature [to the one found in these mice] is found in human cells, it could help researchers to identify which iPS cells to avoid using, and which stand the best chance of producing the desired tissue."

cells on the other hand, which can be standardised and mass produced, would only need one-off approval, making them a more attractive business proposition."[i,41] Using iPSCs may be "uneconomical".[42] In short, there is a need for human embryo research to be continued. In the US, this is evident in the increasing number of new stem cell lines of human embryonic stem cells being approved for research.[j,43]

Perhaps, if the science had started with iPS and body cells rather than with embryos, there might not have been such concern about scientists tampering with the basic elements of human life, and less demand for rigorous regulation. The research and possible new treatments might have appeared more like other medical innovations, to be governed by ethical oversight to ensure that risks are properly evaluated before patients start the treatment, that patients are informed of potential risks and consent voluntarily, that records are kept and adverse events reported, and the like. Such a 'light-touch' regulatory scheme would be supplemented by the incidental legislation on matters like quality assurance that applies to all new products and procedures for use in human health care.

However, even if human embryo research were to be prohibited in the future, the focus on embryos in the early discussion of regulation has left its mark on the discourse regarding stem cell research. Embryos were described as potential persons, and concerns were expressed about the dangers of tampering with potential human life. Even without embryos in the research programs, critics would now ask similar questions about 'a potential new life' that might not have arisen if the issue was how to regulate treatments involving reprogramming a patient's cells to treat the patient's disease.

[i] Dr Thomas Okarma, CEO of Geron Corporation (the first company to get approval from the US FDA to undertake a clinical trial on a stem cell therapy for human patients) — see note 41.

[j] See press release of National Institutes of Health (note 43): 13 human embryonic stem cell (hESC) lines approved for use in NIH-funded research. It has since been reported that four more new embryonic stem cell lines have been approved for use in federally funded research.

Scientifically, questions about 'a potential new life' are not without foundation even if stem cell research does not involve embryos. Recent research suggests that iPSCs can be converted into cells that are very similar to embryonic stem cells, and that those cells may have the potential to develop into an embryo. Scientists have also found that this principle applies not only to iPSCs, but to any other cells of the body. All of the body's cells may potentially be converted into gametes (sperm and eggs) which, at least in theory, could be fertilised to form an embryo. If people object to research on embryos because they are potential persons, the same objection may apply to iPS and body cells.[44] If the 'potentiality' argument is rejected to enable stem cell research to continue, then, for consistency, the objection to human embryo research on the grounds that an embryo is a potential person may also need to be discounted (though the two events, embryo formation and birth of child, are much more closely related when an embryo has already been formed).

These arguments aside, there are more positive reasons to argue that human embryo research should be continued despite the developments in iPSCs and cellular reprogramming. We do not know which type of stem cell research will ultimately be the most successful. As noted earlier, iPSCs may not be as effective as embryonic stem cells in their potential for continued multiplication and sustained stability. At present, the most promising developments seem to be in the area of cellular reprogramming, but many scientists still see benefits in working with embryonic cells.

In addition, there are many reasons to undertake research on embryos in addition to deriving stem cells from the embryos for use in research. The study of early human embryos is vital to understand the process and causes of abnormality in foetal development and to improve techniques in fertility treatment. These types of research can be done only on embryos. Also, research on early embryos is necessary to understand how pluripotent stem cells develop and differentiate into other kinds of cells. Knowledge gained from embryo research has assisted scientists doing research on iPS cells.[45]

5. OTHER POLICY ISSUES: PAYMENT FOR EGGS; HUMAN–ANIMAL HYBRIDS

If human embryo research continues, large numbers of human eggs may be needed,[k,l,46,47] and there is a shortage of human eggs for use in research. This raises the issue of whether monetary payments or other inducements should be permitted for donating human eggs for use in research. In the UK, Canada, and Australia, the tradition in medical research has been that all tissue used in research should be given gratuitously, including human eggs and embryos, and payments are not permitted beyond reasonable expenses,[48–51] such as reimbursement of the donors' medical expenses and compensation for loss of earnings due to the donation. Similarly, European countries disapprove of the commercialisation of or obtaining financial gains from the donation of human reproductive materials. However, in the UK, the policy of the Human Fertilisation and Embryology Authority (HFEA) on 'egg sharing' modifies the general approach of altruism to some extent, as women who are prepared to donate some of their eggs for research may gain accelerated access to fertility treatment programs. The US goes further. There is no federal legislation governing the sale of human eggs, and they may be sold for a 'fair price' for use in fertility programs[m] and in research. The meaning of 'a fair price' is open to interpretation.[52] The state of New York legislated to allow federally funded researchers to pay women for donating their eggs for research.[53]

Even if the law permits women to sell their eggs for research, it is unlikely that there will be a large number available for research. Egg

---

[k] According to an Australian newspaper report, 227 sheep eggs were used to create Dolly and 304 monkey eggs to create the first two primate ES cell lines (see note 46).

[l] In 2009, a Chinese research team reported the creation of patient-specific embryonic stem cells in which "a total of 135 oocytes were obtained from 12 healthy donors (30–35 years)" (see note 47).

[m] Some states, however, have legislative restrictions on the sale of eggs for use in research. California, for example, prohibits the purchase or sale of an ovum, zygote, embryo, or foetus for the purpose of cloning human beings.

donation is invasive and may involve risks that we do not yet know of. However, there may be an alternative. It may be possible to 'incubate' the nucleus of a human cell in an enucleated animal egg in order to produce embryonic stem cells that are almost entirely human. This is currently prohibited in Australia,[54] Canada,[55] and in many European and other countries. It is not currently banned in the US (the Human–Animal Hybrid Prohibition Bill was introduced in 2008 but has not been passed), but federal funding is not permitted for this research in the US.

In the UK, on the other hand, it is lawful. The first human–animal embryo was created in 2008 by scientists at Newcastle University under a licence from the HFEA.[56] The validity of such a licence was confirmed in 2008 when the Human Fertilisation and Embryology Act 1990 was amended by Parliament.[57–59] Since then, two more licences have been granted, and the first human–animal embryo has been formed, though no stem cells have yet been derived from it. However, despite having licences to do this research, scientists have not been able to get research grants to do it.[60] Perhaps one reason is that the funding bodies and their reviewers do not believe that the proposed research will be successful, or they consider that other research will be more productive. In either event, this is perhaps an example of research finding its own 'level', in accordance with my argument, without the need for it to be banned by legislation.

6. PITFALLS IN AMENDING LEGISLATION: THE EXAMPLE OF NEW RESEARCH ON MITOCHONDRIAL DISEASE

There are other reasons to be cautious in enacting legislation and, *a fortiori*, in amending legislation to deal with scientific developments and changing attitudes. Regulating in an area of rapid change, like stem cell technology, inevitably leads to gaps and inconsistencies in policy. Consider the recently reported research in Newcastle UK which may lead to a future treatment for mitochondrial disease.[61] The Newcastle team formed a human embryo by removing healthy DNA from a fertilised egg formed from a couple's gametes (sperm and egg), and placed that DNA in an egg from a donor with healthy mitochondria after removing the nucleus from that egg. The resultant embryo contained DNA from three people, but adding the healthy mitochondria from the egg donor could lead to the birth of a child free of the mother's mitochondrial disease. Now consider this embryo with the DNA from three people. Critics will argue that it is wrong to create such an embryo — we should not interfere with Nature; we do not know the potential consequences of this type of intervention; couples should not seek 'designer babies' but accept and love the children they are given. However, whatever the ethical sensitivity about creating such an embryo it should not matter how the embryo is formed. The scientific and ethical point is that the embryo contains DNA from three people. Yet, in Australia, this procedure would be an offence punishable by up to 15 years imprisonment if the embryo were to be formed by fertilising a human egg with human sperm (Prohibition of Human Cloning for Reproduction Act 2002 (Cth), s 13), but not if it were formed by another process, such as somatic cell nuclear transfer (SCNT), performed under licence (s 23(a) of that Act; Research involving Human Embryos Act 2002 (Cth), s 20). The latter provisions were added to the original legislation when it was amended in 2005. When the legislation was first passed in 2002, the only way to form an embryo was by fertilising an egg with sperm. No one anticipated the possibility of SCNT for this purpose.

There are many other examples of provisions that seem to be inconsistent. In Australia, it is lawful to create embryos for research by SCNT, but not by fertilising a human egg with human sperm (though that is lawful in the UK).[n] However, if eggs are donated for research and sperm is donated for research, why should a researcher not be permitted to fertilise the eggs for research, subject to obtaining a licence and complying with other regulatory requirements? Also, generating human embryos for research may

---

[n] It is true that a 'sperm–egg embryo' is made up of genetic material from both parents equally whereas a 'SCNT embryo' has genetic material mostly from the person whose body cell was used to form the embryo, with only a small amount of mitochondrial DNA from the donated egg used to incubate that genetic material. However, that difference does not seem morally significant where the donors are not intentionally trying to combine their DNA to have a child. Where both donors are in an IVF program, their view of an embryo formed from their own gametes may be different from the view of donors of eggs and sperm separately.

be better than using donated embryos, from a scientific and ethical perspective. Cells obtained from these embryos would not have been frozen, and donors of the sperm and eggs would know in advance that the embryos would be created to be used for research,[62] rather than facing this possibility once their embryos have been declared 'excess' in a fertility treatment program.

7. THE HEEREY COMMITTEE'S RECOMMENDATIONS

In Australia, the Heerey Committee's report was tabled in the federal Parliament on 7 July 2011.[63] It recommended that the basic structure of the Australian legislation should remain the same. It acknowledged the progress that has been made with iPS cells but it accepted that human embryo research is still justifiable and said that it should continue to be allowed subject to licence and ethical oversight. This included research involving human embryos created for research by somatic cell nuclear transfer (SCNT) as well as embryos donated from fertility treatment programs. However, the Committee sounded a note of caution about the use of embryos created by SCNT for research. Noting the lack of progress in SCNT research in animals as well as humans, it said that the Licensing Committee should take account of that lack of progress when considering future applications to create embryos by SCNT for research. The Committee should pay particular attention to its current statutory duty to consider in any licence application 'the likelihood of significant advance in knowledge or improvement in technologies for treatment as a result of the use of excess ART embryos or human eggs, or the creation or use of other embryos, proposed in the application, which could not reasonably be achieved by other means' (s 21(4) of the Research Involving Human Embryos Act 2002).

In other recommendations, the Committee said that payment for gametes should not be allowed beyond the payment of 'reasonable expenses', as under the current law. Also, the legislation should not be amended to assist research into mitochondrial disease by allowing an embryo to be created with DNA from more than two people by fertilisation of a human egg and sperm (the Committee considered the research in the UK is not sufficiently advanced at this stage and other research should be done before allowing it in Australia with human genetic material).

The Committee mentioned the creation of human animal embryos for research and did not recommend that this should be allowed. However, it devoted more attention to the use in research of precursor cells and IVD gametes, which it said could be defined as 'human sperm or eggs derived from precursor cells or by *in vitro* means': Recommendation 12)'. The Committee noted that the possible use of this technology in human reproduction is to date theoretical and that there has been little discussion of the underlying issues in the wider community. It said that the reproductive use of human IVD gametes should not be permitted at the present time but it supported research using precursor cells and human IVD gametes under licence to advance scientific knowledge.

8. CONCLUSION

This paper has outlined some of the principles in the early regulatory structure concerning human embryo research, which are common to many countries. Because a human embryo is the first stage of human life, there has been widespread concern about scientists tampering with human life and potential misuse of the technology. This has led to legislative controls which involve criminal penalties even if phrased in the terms of a licensing system. However, the most recent scientific developments in stem cell technology have enabled stem cells to be derived without the destruction of embryos (iPSCs). Also, ordinary body cells have been reprogrammed into stem cells which can be used in human health care. Already some successful treatments have been reported and clinical trials have commenced.

It has been suggested that the development of iPSCs and cellular reprogramming may not resolve all the ethical issues associated with human embryo research, especially if the focus of the discourse continues to be the moral significance of undertaking destructive research on 'potential persons'. Given the problems that arise when legislation is amended, with the inevitable risk of creating gaps and inconsistencies, perhaps we should halt further steps to change the regulatory structure. Australian scientists do not seem to be clamouring to extend the research that they are currently undertaking, at least at the moment, and the Heerey Committee saw no need for legislative changes apart from some minor administrative adjustments. Groups of scientists at stem cell conferences, whom I have informally asked whether the current law is restricting their research activities, have said that it is not. They have all said that the law is generally operating well, and they are not seeking changes. There seems to be no stem cell research that they want to do that the legislation prevents them from undertaking. Most seem to be happy to leave the current legislation as it is. Minimal legislative change would also avoid the costs of amending the legislation, both federally and in the states and territories, and perhaps ending up with provisions that are more restrictive than those that exist at present.

REFERENCES

1. Australian Government. (2005) *Legislation Review: Prohibition of Human Cloning Act 2002 and the Research Involving Human Embryos Act 2002* (Reports, Canberra, December 2005). Available at http://pandora.nla.gov.au/pan/63190/20060912-0000/www.lockhartreview.com.au/reports.html (Accessed 5 Aug 2011).
2. Prohibition of Human Cloning for Reproduction Act 2002 (Australia).
3. Research Involving Human Embryos Act 2002 (Australia).
4. See https://legislationreview.nhmrc.gov.au/2010-legislation-review (Accessed 5 Aug 2011).
5. Steinberg A. (1 June 2010) *Jewish Medical Ethics: May Humans Play God?* (lecture), Oxford.
6. Henderson M. (24 March 2008) Cloned Cells Bring Hope of Therapy for Parkinson's Disease. *The Times Online*. Available at www.timesonline.co.uk/tol/news/science/article3607659.ece (Accessed 5 Aug 2011).
7. Madeleine BLL. (22 June 2005) Embryonic Stem Cell Research: Accepting the Knowledge and Applying It to Our Lives. *Japan Inc. Communications*.

Available at http://www.thefreelibrary.com/Embryonic+stem+cell+research:+accepting+the+knowledge+and+applying+it...-a0134293286 (Accessed 5 Aug 2011).

8. Hanna J, *et al.* (2007) Treatment of Sickle Cell Anemia Mouse Model with iPS Cells Generated from Autologous Skin. *Science* **318:** 1920–23.
9. Hanna J, *et al.* (2007) Treatment of Sickle Cell Anemia Mouse Model with iPS Cells Generated from Autologous Skin. *Science* **318:** 1920–23.
10. Hanna J, *et al.* (2007) Treatment of Sickle Cell Anemia Mouse Model with iPS Cells Generated from Autologous Skin. *Science* **318:** 1920–23.
11. Midgley S. (10 May 2008) Spine Injuries May Be Repaired by a Nose. *The Times Online.* Available at http://business.timesonline.co.uk/tol/business/specials/stemcell_research/article3904111.ece (Accessed 5 Aug 2011).
12. Science Timeline, Brits at Their Best. Available at http://www.britsattheirbest.com/ingenious/ii_21st_century.htm (Accessed 5 Aug 2011).
13. Anon. (20 May 2008) Veterinarians Achieve Success with Adult Stem Cell Therapy in Animals. *Stem Cell Institute.* Available at http://www.cellmedicine.com/vet-stem-cells/ (Accessed 5 Aug 2011).
14. Normile D. (4 Aug 2011) Sperm Made (Mostly) in a Dish Produce Normal Mice Available at: http://news.sciencemag.org/sciencenow/2011/08/sperm-made-mostly-in-a-dish-produce.html?ref=hp (Accessed 5 Aug 2011); Hinxton Group. The Science, Ethics and Polity Challenges of Pluripotent Stem-Cell Derived Gametes, April 2008. Available at http://www.hinxton-group.org/HinxtonConsensus_April2008.doc (Accessed 5 Aug 2011).
15. Josephs L. (2 June 2010) Costa Rica Shuts Stem Cell Clinic. *Reuters.* Available at http://www.reuters.com/article/2010/06/02/us-costarica-stemcells-idUSTRE6516UR20100602 (Accessed 5 Aug 2011).
16. Lako M, Armstrong L, Stojkovic M. (2010) Induced Pluripotent Stem Cells: It Looks Simple but Can Looks Deceive? *Stem Cells Express* **28(5):** 845–850.
17. Sipp D. (2010) Challenges in the Clinical Application of Induced Pluripotent Stem Cells. *Stem Cell Research & Therapy* **1:** 9.
18. Carney C. (24 February 2007) Stem Cell Success for Spinal Injury in India. *Wired.* Available at http://www.wired.com/bodyhack/2007/02/stem_cell_succe/ (Accessed 5 Aug 2011).
19. McArthur G. (30 May 2008) Miracle Man Says Embryonic Stem-Cell Therapy Works. *Herald Sun.* Available at http://www.heraldsun.com.au/ news/victoria/miracle-man-takes-steps-to-recovery/story-e6frf7kx-1111116481636 (Accessed 5 Aug 2011).
20. Falco M. (21 Jan 2010) First U.S. Stem Cells Transplanted into Spinal Cord. *CNN.* Available at http://edition.cnn.com/2010/HEALTH/01/21/stem.cell.

spine/index.html?utm_source=Full+List&utm_campaign=73859fdf92-SCOPE1_14_2010&utm_medium=email (Accessed 5 Aug 2011).
21. Cressey D. (10 February 2010) Stem Cell Stroke Trial Gets Final Approval in UK. The Great Beyond. *Nature Blog*. Available at http://blogs.nature.com/news/2010/02/stem_cell_stroke_trial_gets_fi.html (Accessed 5 Aug 2011).
22. Lister S. (30 January 2009) Stem Cell Therapy Reduces Symptoms of Multiple Sclerosis. *The Times Online*. Available at http://www.timesonline.co.uk/tol/life_and_style/health/article5614644.ece (Accessed 5 Aug 2011).
23. Burt RK, *et al.* (2009) Autologous Non-Myeloablative Haemopoietic Stem Cell Transplantation in Relapsing-Remitting Multiple Sclerosis: A Phase I/II Study. *The Lancet Neurology* **8:** 244–253 (This describes in detail the above-mentioned study).
24. Karolinska Institutet. (2 February 2010) Stem Cells Rescue Nerve Cells by Direct Contact. *Science Daily*. Available at http://www.sciencedaily.com/releases/2010/02/100201171754.htm (Accessed 5 Aug 2011).
25. Templeton SK. (14 Sept 2009) Stem Cells Believed to Repair Damaged Hearts. *The Australian*. Available at http://www.theaustralian.com.au/news/world/stem-cells-believed-to-repair-damaged-hearts/story-e6frg6so-1225772469533 (Accessed 5 Aug 2011).
26. Henderson M. (23 December 2009) Stem Cell Eye Treatment Gives Victim of Fight his Sight Back. *The Times Online*. Available at http://www.timesonline.co.uk/tol/news/uk/health/article6965043.ece (Accessed 5 Aug 2011).
27. Henderson M. (23 December 2009) Stem Cell Eye Treatment Gives Victim of Fight his Sight Back. The Times Online (quoting Dr Sajjad Ahmad, co-leader of the eye treatment trial in Newcastle, UK). Available at http://www.timesonline.co.uk/tol/news/uk/health/article6965043.ece (Accessed 5 Aug 2011).
28. Anon. (21 October 2009) Transplanted Tissue Improves Vision: Study Shows Enhanced Visual Acuity. *Science Daily*. Available at http://www.sciencedaily.com/releases/2009/10/091021014628.htm (Accessed 5 Aug 2011).
29. Akst J. (19 November 2009) 2nd Human hESC trial? *The Scientist*. Available at http://classic.the-scientist.com/blog/display/56155/ (Accessed 5 Aug 2011). Sheridan K. (22 Nov 2010) Stem Cells May Help Treat Blindness. *The Age*. Available at http://news.theage.com.au/breaking-news-world/stem-cells-may-help-treat-blindness-20101122-184bw.html (Accessed 5 Aug 2011).
30. Gray R. (27 Apr 2009) The Miracle Stem Cell Cures Made in Britain. *The Telegraph*. Available at http://www.telegraph.co.uk/technology/5232182/The-miracle-stem-cell-cures-made-in-Britain.html (Accessed 5 Aug 2011).

31. Anon. (2008) A Stem Cell Success Story — UK Stem Cell Biologists Help to Deliver a New Bronchus for Claudia Castillo. *UK National Stem Cell Network Newsletter*, Winter: 8. Available at http://www.uknscn.org/downloads/newsletter_winter08.pdf (Accessed 5 Aug 2011).
32. See also a later case in the UK: Laurance J. (2010) British Boy Receives Trachea Transplant Built with his Own Stem Cells. *British medical Journal* **340:** 1633.
33. Harris D. (12 October 2009) Groundbreaking Stem Cell Surgery Gives Boy New Cheekbones. *Good Morning America*. Available at http://abcnews.go.com/GMA/OnCall/experimental-treatment-boy-cheekbones/Story?id=8804636&page (Accessed 5 Aug 2011).
34. California Institute for Regenerative Medicine (CIRM). (11 March 2010) CIRM Allocates $50 Million for Stem Cell Therapy Development, a Boost for the State's Growing Stem Cell Industry (press release). Available at http://www.cirm.ca.gov/PressRelease_031110 (Accessed 5 Aug 2011).
35. Banjo S. (14 May 2010) Hedge-Fund Founder Bolsters Stem-Cell Research With $27 Million Gift. *Wall Street Journal*. Available at http://online.wsj.com/article/SB10001424052748704635204575242444135532502.html (Accessed 5 Aug 2011).
36. Winslow R, Mundy A. (23 January 2009) First Embryonic Stem-cell Trial Gets Approval from the FDA. *Wall Street Journal*. Available at http://online.wsj.com/article/SB123268485825709415.html (Accessed 5 Aug 2011). Sheridan K. (22 Nov 2010) Stem Cells May Help Treat Blindness. *The Age*. Available at http://news.theage.com.au/breaking-news-world/stem-cells-may-help-treat-blindness-20101122-184bw.html (Accessed 5 Aug 2011).
37. Akst J. (19 November 2009) 2nd Human hESC trial? *The Scientist*. Available at http://classic.the-scientist.com/blog/display/56155/ (Accessed 5 Aug 2011).
38. American Association for the Advancement of Science. (18 December 2008) Cellular Reprogramming Leads Science List of Top 10 2008 Breakthroughs (press release). Available at http://www.aaas.org/news/releases/2008/1218breakthrough.shtml (Accessed 5 Aug 2011).
39. Panizzo R. (12 April 2010) DNA Difference Between Stem Cell Types Found. *BioNews*. Available at http://www.bionews.org.uk/page_57804.asp (Accessed 5 Aug 2011).
40. Dolgin E. (2010) Gene Flaw Found in Induced Stem Cells. *Nature* **464:** 663.
41. Pryor S. (22 February 2010) Fresh Fears over the Promise of iPS Cells. *BioNews*. Available at http://www.bionews.org.uk/page_54791.asp (Accessed 5 Aug 2011).

42. Henderson M. (17 February 2010) Medical Potential of iPS Stem Cells Exaggerated says World Authority. *The Times Online*. Available at http://www.timesonline.co.uk/tol/news/science/medicine/article7029447.ece (Accessed 5 Aug 2011).
43. National Institutes of Health. (2 December 2009) First Human Embryonic Stem Cell Lines Approved for Use Under New NIH Guidelines. Available at http://www.nih.gov/news/health/dec2009/od-02.htm (Accessed 5 Aug 2011).
44. Devolder K. (2009) To Be, or Not to Be? *EMBO Reports* **10(12):** 1285–1877. Here, Devolder asks whether induced pluripotent stem cells are potential babies, and whether it matters.
45. Hall K. (3 February 2009) Japanese Stem Cell Scientist Announces Promising Results. *Business Week*. Available at http://www.businessweek.com/blogs/eyeonasia/archives/2009/02/japanese_stem_cell_scientist_announces_promising_results.html (Accessed 5 Aug 2011).
46. Martin J. (17 January 2008) Amen to Death of Embryo Research. *The Australian*. Available at http://www.theaustralian.com.au/news/opinion/amen-to-death-of-embryo-research/story-e6frg7ef-1111115333189 (Accessed 5 Aug 2011).
47. Anon. (3 February 2009) Chinese Researchers Make Cloned Human Blastocysts. *Cell News*. Available at http://cellnews-blog.blogspot.com/ 2009/02/chinese-researchers-make-cloned-human.html (Accessed 5 Aug 2011).
48. Assisted Human Reproduction Act S.C. 2004, c. 2, §§ 6, 7, 12 (Canada).
49. Prohibition of Human Cloning for Reproduction Act, 2002, No. 144, § 21 (Australia).
50. Canadian Institutes of Health Research. (2010) *Updated Guidelines for Human Pluripotent Stem Cell Research*, § 4.0. Available at http://www.cihr-irsc.gc.ca/e/42071.html (Accessed 5 Aug 2011)
51. Mills P. (2004) *Human Fertilisation and Embryology Authority, Seed Review Consultation Proposals* (ELC (09/04) 02). Available at http://www.hfea.gov.uk/docs/ELC_SEED_Sept04.pdf (Accessed 5 Aug 2011).
52. Isasi RM, Knoppers BM. (2007) Monetary Payments for the Procurement of Oocytes for Stem Cell Research: In Search of Ethical and Political Consistency. *Stem Cell Research* **1:** 37–44, pp. 37 & 39.
53. Siemaszko C. (26 June 2009) $10,000 is an Egg-cellent Price, Says Stem Cell Panel. *Daily News*. Available at http://articles.nydailynews.com/2009–06–26/news/17925165_1_stem-cell-religious-groups (Accessed 5 Aug 2011).
54. Prohibition of Human Cloning for Reproduction Act 2002 (Cth) §§ 8(1), 17, 18.
55. Assisted Human Reproduction Act 2004 (Canada), §§ 5(1)(i), (j).

56. Henderson M. (2 April 2009) We Have Created Human–Animal Embryos Already, Say British Team. *The Times Online.* Available at http://www.timesonline.co.uk/tol/life_and_style/health/article3663033.ece (Accessed 5 Aug 2011).
57. Human Fertilization & Embryology Authority. (13 November 2008) HFEA Chair Welcomes Royal Assent for HFE Act (press release). Available at http://www.hfea.gov.uk/371.html (Accessed 5 Aug 2011).
58. Lovell-Badge R. (2008) The Regulation of Human Embryo and Stem-Cell Research in the United Kingdom. *Nature Reviews Molecular Cell Biology* **9:** 998–1003, p. 1001. Note that creating a 'true hybrid' by fertilising a human egg with animal sperm or vice-versa is unlawful.
59. Batty D. (17 May 2007) Hybrid Embryos Get Go-Ahead. *The Guardian.* Available at http://www.guardian.co.uk/science/2007/may/17/businessofresearch.medicineandhealth (Accessed 5 Aug 2011).
60. Connor S. (5 October 2009) Vital Embryo Research Driven Out of Britain. *The Independent.* Available at http://www.independent.co.uk/news/science/vital-embryo-research-driven-out-of-britain-1797821.html (Accessed 5 Aug 2011).
61. Turnbull DM, Tuppen HA, Greggains GD, *et al.* (2010) Pronuclear Transfer in Human Embryos to Prevent Transmission of Mitochondrial DNA Disease. *Nature* **465:** 82–85.
62. Friend T. (11 July 2001) Group Creates Embryos Specifically for Research. *USA Today.* Available at http://cmbi.bjmu.edu.cn/news/0107/89.htm (Accessed 5 Aug 2011).
63. Report of the independent Review of the Prohibition of Human Cloning for Reproduction Act 2002 and the Research Involving Human Embryos Act 2002. Available at: https://legislationreview.nhmrc.gov.au/sites/default/files/legisiation_review_reports.pdf (Accessed 5 Aug 2011).

5

Human Embryos in Stem Cell Research: Property and Recompense

Sarah Devaney

The Court of Appeal of England and Wales ruled in the case of *Yearworth and Others v North Bristol NHS Trust* that sperm stored by six men for their future reproductive use was the property of its progenitors. This chapter explores whether the status of property which was bestowed upon sperm intended for reproductive use can equally be applied to human embryos created through *in vitro* fertilisation treatment and intended for use in stem cell research. It will conclude that, while embryos can have certain values in a reproductive setting, once the individuals whose gametes have been combined to form the embryo have decided that they should be made available for research purposes, the law allows for them to assume the status of property, for which appropriate recompense should be made. A number of concerns which might be raised by this conclusion are discussed, leading to the argument that none of them constitute compelling reasons to deny property status to embryos used in stem cell research. Rather, they serve to confirm that appropriate regulation which is fair to all participants in the stem cell research endeavour may contribute to the realisation of the hopes for this technology.

Centre for Social Ethics & Policy, School of Law, University of Manchester, Oxford Rd., Manchester, M13 9PL, United Kingdom. Email: sarah.devaney@manchester.ac.uk

1. INTRODUCTION

On 4 February 2009, the Court of Appeal of England and Wales ruled in the case of *Yearworth and Others v North Bristol NHS Trust*[1] that sperm stored by six men for their future reproductive use was the property of its progenitors. This finding resulted in awards of compensation to those men after insufficient care was taken by the clinic storing the samples, which were irretrievably damaged as a result. The ruling has implications for those making their tissues available for use in stem cell (SC) research. In the world of biotechnology, waste in the form of unwanted bodily tissue can be transformed into wealth.[2] It is also hoped that in the SC arena, such unwanted tissues can be transformed into health through the development of therapies for serious diseases. In order to achieve this, SC researchers require the provision of a variety of types of human tissues which may be made available by a number of categories of human providers. This chapter will explore whether the status of property which was bestowed upon sperm intended for reproductive use can equally be applied to human embryos created through *in vitro* fertilisation (IVF) treatment and intended for use in SC research.[a]

There are three reasons for the selection of this type of tissue. First, in addition to the possibility of creating SCs by reprogramming adult cells,[3] the most potent and adaptable type of SCs, pluripotent SCs, can be found in the inner cell mass of the five- to seven-day-old embryo, known as a blastocyst. As a result, embryos represent a major research focus in SC science. Not all embryos made available to SC research will be of sufficient quality to yield SCs: First, because the highest quality embryos are chosen for implantation,[b,4] and second, because not all of those which are

[a] Analogous arguments to those made in this chapter could be made in relation to the property status of embryos within the reproductive context. However, the focus of this piece is the property status of embryos within the research context, leaving the question of whether an embryo is property in a reproductive context open.

[b] Although a January 2008 study reported that even those embryos which would not be clinically suitable for reproductive purposes could yield SC lines at a rate similar to frozen embryos — see note 4.

provided after cryopreservation will survive the freezing, thawing, and culturing process.[5] Nevertheless, improvements in blastocyst-culture techniques,[6] together with the recommendation by the Human Fertilisation and Embryology Authority (HFEA)[c] that fertility clinics should aim to implant only one embryo during IVF,[d,7–9] mean that significant numbers of IVF embryos will not be used for reproductive purposes and may be provided to researchers. Although it is hoped that, as a result of the development of a SC line bank providing a research resource, the need for new embryos in research will decline over time,[10] few lines have to date been banked and made available for access to researchers.[11] As a result of these factors, IVF embryos are likely to remain a significant source of research material for SC research.[e]

The second reason for selecting embryos as the focus of this analysis is due to the contentious nature of debates about their status. Objections to the classification of embryos as property are not uncommon[12] as the nature of embryos within SC research in the UK is "complicated, paradoxical and over-determined: they are extremely valuable but cannot be sold, and are instead often categorized as 'gifts'".[13] The way in which they are valued is connected to the fact that they could become a born human being given the right circumstances and significant amounts of luck. However, they also have an important value in SC research on the basis of the powerful cells they contain. This discussion will not attempt to resolve the intractable debate about the *moral* status of the embryo. Rather, it is based on the premise that there is an irresolvable chasm between the extremes of this debate, and that a regulatory framework must function

[c] The UK regulatory authority which oversees the use of embryos in assisted reproduction and research through the issuing of licenses to those centres undertaking this work.

[d] Note that the HFEA's review, titled "The Best Possible Start to Life", which commenced in April 2007, considered the issue of whether the number of embryos permitted to be transferred at any one time should be reduced in light of the risks posed to mothers and babies by multiple pregnancies. As a result, the HFEA has implemented a policy which aims to reduce the numbers of multiple births arising out of assisted conception techniques to 10% over a number of years. See notes 8 and 9.

[e] The arguments made here can equally be applied to embryos created through Somatic Cell Nuclear Transfer or 'therapeutic cloning'.

appropriately despite this. To do so, it must be clear about its position on the *legal* status of these entities where they are intended to be used in SC research. Finally, if convincing arguments can be made that the legal status of property can properly be applied to this most contentious form of tissue used in the SC context, the application of this status to all other forms of tissue thus used, such as ova or skin cells, should follow more easily.

In Section 2 of this chapter, the question of whether embryos used in SC research can be classed as property will be examined. To this end, the nature of the embryo in this context and the role and thus entitlements of its providers will be analysed. The conclusion will be drawn that where the progenitors of embryos no longer intend to use them for reproductive purposes, instead deciding to make them available for SC research, those embryos are property for which their providers should be fairly recompensed. In Section 3, three particular challenges for the regulatory sphere in protecting the interests of embryo providers in this context will be dealt with: protecting patients from conflicts of interest on the part of their treating clinicians; countering allegations of commodification; and determining what should happen where tissue providers fail to indicate or agree on what should happen to their embryos. It will be argued that while these concerns require regulatory oversight, where such oversight is in place, they do not constitute sufficient grounds to exclude embryos from the category of property.

2. IVF EMBRYOS AND CONCEPTS OF PROPERTY

In this section, the argument will be made that those legal provisions which permit the progenitors of IVF embryos to decide whether these tissues are used in research represent an implicit acknowledgement of their property rights over them. An examination of the views of embryo progenitors about these tissues in the research context will be undertaken, leading to the conclusion that the law's implicit acknowledgement of their status as owners in no way conflicts with the progenitors' views about those embryos which are donated to research. This leads to a consideration of the nature and level of recompense to which such progenitors are entitled.

2.1 The Law's Approach to IVF Embryo Progenitors

During IVF treatment, a woman will receive drugs to stimulate superovulation so that as many ova as possible can be harvested from her body before being retrieved surgically and combined with the sperm of her partner or a donor to create embryos. As a result of the HFEA's policy of limiting the number of embryos which can be implanted in a woman in a cycle (up to a maximum of three depending on the circumstances, but with the intention that single embryo transfer will occur in most cases),[14] often more embryos are created than can legally be placed immediately in a woman. The primary legal provision forming the cornerstone of permissible acts in relation to embryonic SC research, the Human Fertilisation and Embryology Act 1990 (as amended) (the 1990 Act), sets out what may be done with the supernumerary of 'spare' embryos. The woman, or couple, can choose to donate fresh spare embryos to research if they are invited to do so during the course of their fertility treatment. If the funds are available, such embryos can also be frozen, or 'cryopreserved', by the clinic providing the treatment. These can be stored for up to 10 years,[f,15] and if in storage at the end of that period will be allowed to perish[g,16] in the absence of any alternative direction by their progenitors. Once the family of the woman or couple is complete, or they are unable or unwilling to undergo further fertility treatment for financial or other reasons, they are legally entitled to make a decision about the fate of any spare embryos held in storage. In the research context therefore, the legal framework recognises the intrinsic value of embryos by placing restrictions on

[f] This period may be extended by regulations, see for example the Human Fertilisation and Embryology (Statutory Storage Period for Embryos and Gametes) Regulations 2009/1582, which allows for an extension in relation to patients with premature infertility.

[g] Schedule 3ZA of the 1990 Act provides at paragraph 2 that counselling must be offered as a condition of a licence for treatment where treatment services involving the use of an IVF embryo are being offered. Such counselling will cover the implications of storing embryos — See HFEA, Code of Practice, Guidance note 4.2(i).

their use in research, but acknowledges their instrumental value and facilitates certain attempts to realise that value, within the realms of the progenitors' wishes. In doing so, it acknowledges and respects a relationship of property between the progenitors and their embryos.

The assertion that this relationship is one of property can be explored by testing the law against Honoré's bundle of rights, duties and responsibilities which can jointly and severally indicate such a relationship.[17] This bundle consists of:

> [T]he right to possess, the right to use, the right to manage, the right to the income of the thing, the right to the capital, the right to security, the rights or incidents of transmissibility and absence of term, the duty to prevent harm, liability to execution, and the incident of residuarity.[18]

One of the most fundamental legal provisions in this context is that:

> [A]n embryo the creation of which was brought about *in vitro* must not be used for any purpose unless there is an effective consent by each relevant person in relation to the embryo to the use for that purpose of the embryo and the embryo is used in accordance with those consents.[h,19]

Paragraph 4 of Schedule 3 of the 1990 Act provides for consent to the use or storage of an embryo to be varied or withdrawn, but this cannot be done once the embryo has been used in treatment services or research. This is defined as being "as soon as they are under the control of researchers or trainers/trainees and are being cultured for use in research or training".[20] Embryos may not be created using an individual's gametes or cells, or used, unless an effective consent[i,21] by that individual is in place for the embryo being put to one of the legally permitted uses, in this context, research.[22]

With these provisions, the legislation allows the progenitors to help to shape the legal status of embryos in practical terms. Bestowing the ability

---

[h] Relevant persons are defined in paragraph 6(3E) and include those whose gametes or cells were used leading to the creation of an embryo or human admixed embryo.

[i] That is, a consent "that has not been withdrawn" — see note 21.

to determine the fate of spare embryos upon gamete providers grants them property-type powers over them.[23] At the point of provision to research, although the embryos do not lose the intrinsic potential of being able to develop into a born human being given the right conditions, this will never take place as the gatekeepers to this (the gamete providers) have not provided a legally valid consent for this to occur. As a result, they may be used as a research raw material, that is, they acquire the new legal status which the progenitors have given permission for. Now, the embryo's special property of containing SCs has become the focus of their use. The licensing system overseen by the HFEA facilitates a situation whereby, where they are agreed on the course of action to be pursued, the progenitors have joint control over the fate of their embryos,[j,24] and together can choose to transfer that control to SC researchers,[k,25] akin to the 'right to manage' strand of the property bundle.[26]

Further indications of the relationship of owner and property between gamete providers and their embryos can be found in the information and advice required by law to be given to those considering provision of their embryos to SC research. In addition to technical requirements for the form of consent required in relation to the use of embryos,[27] such a consent must specify which of the legally permitted uses (treatment, training of clinicians or research) to which the embryos may be put and may specify conditions to that consent.[28] Appropriate relevant information must be given to the provider in advance of their giving consent, together with the opportunity to receive counselling.[29]

The HFEA provides in licence condition R19 that people making gametes or cells available which will be used to create research embryos

---

[j] Where a gamete provider withdraws consent for the storage of an embryo intended for treatment purposes, the other gamete provider will have to be notified, and unless they too consent to the embryo's destruction, the embryo will not be destroyed for a year following notification of withdrawal of the consent. If no joint decision is made about the fate of the embryo, the default position is that it will be destroyed — see note 24.

[k] The law does permit embryo progenitors to continue to assert some control over them by allowing them specify conditions in relation to research carried out on their embryos — see note 25.

should be given information about the nature of the research project. They must be informed of a number of issues including that donation to it will not affect any treatment they are undergoing; that they can vary or withdraw consent until the embryos are used in the research; whether the embryos will be reversibly or irreversibly anonymised, and the implications of this; whether they will receive any information in the future; and "how the research is funded, including any benefit which will accrue to the researchers and/or their departments".[30]

In the context of SC research specifically, HFEA licence condition R20 provides that additional information must be given where a person makes gametes or cells available for use in research deriving embryonic SC lines. These include terms which are directed towards restricting the ownership entitlements available to embryo providers such as:

(a) that once an embryo or human admixed embryo has been used in the project of research they will have no control over any future use of the embryonic cells or any stem cells derived …
(d) that the stem cells/lines may be used for commercial purposes, but that they will not benefit financially from this, and
(e) that any stem cells/lines derived or discoveries made using them, could be patented, but that they will not benefit financially from this.[31]

There is a concern that clinics will be able to exert undue influence over patients' decisions about the fate of their embryos in light of the gratitude that they are likely to feel towards their clinician where the medical interventions to assist in reproduction have succeeded in their aim.[32] In order to alleviate this concern, clinics "should ensure that clinical and research roles are separated. Individuals involved in advising patients when making clinical decisions about their licensed treatment should not be involved in research or training that patients are considering donating to".[33] Suitably qualified individuals with knowledge of the proposed research, its risk and implications, should discuss the project with the donor,[34] with a separation of clinical and research roles being made clear to the potential donor.

A great deal of care is therefore taken to ensure that significant amounts of information are given to embryo progenitors in order to enable them to make a decision about whether or not to make their embryos available for research. This bestowal of 'trespassory control'[35] over the embryos indicates an acknowledgement by legislators and the regulatory agency of their rights to possess (defined as having "exclusive physical control of the thing, or to have such control as the nature of the thing admits"[36]) and manage this form of property as owners of it. In *Yearworth*,[37] it was the element of control bestowed upon the progenitors of tissue outside of the body that convinced the Court of Appeal that sperm was property:

> [B]y its provisions for consent, the [1990] Act assiduously preserves the ability of the men to direct that the sperm be *not* used in a certain way: their negative control over its use remains absolute.[38]

That such provisions, in which TPs can direct the use of their tissues, are indicative of a property relationship with cells, forms the basis of a compelling argument that embryos are a form of property. However, their progenitors are unjustifiably excluded from accessing the income and/or capital as part of the property bundle to which they are entitled.[l,39]

In the UK, the decision about the use to which an embryo can be put is placed in the hands of its progenitors. This is within a legal framework shaped by a policy which permits research upon the embryo to take place. In a legal context therefore, the research embryo can validly be viewed as something other than a potential human being. By the time a decision is made to provide embryos for research, the reproductive purpose for which they were created has, for whatever reason, become redundant. At the point embryo progenitors make their decision, their intention helps to shape the embryo's *legal* status, indicating the context within which its value (whether reproductive, therapeutic, or other) should be determined. On attaining the legal status of research raw materials, embryos fall into a category of resources used in SC research, such as laboratory equipment, for which payment is made. Having accepted that embryos may be used in

[l] S12(e) 1990 Act provides that "no money or other benefit shall be given or received in respect of any supply of gametes, embryos or human admixed embryos unless authorised by directions".

research, and given the financial rewards which researchers and biotechnology companies might reap from their use, any claims that the regulatory system enshrines fairness to all its subjects become less credible in light of the failure to permit payment for this form of property. In the following section, the role and views of embryo providers will be analysed in further detail to clarify why it is reasonable to argue that their relationship with embryos on designation for research purposes becomes that of an owner and their property.

2.2 The Nature of the Embryo in SC Research — Progenitors' Views

> The embryo is a deeply, perhaps irretrievably ambiguous entity, one that defies classification and slips seamlessly between moral and biological categories.[40]

It is important that the implicit recognition in law of the property relationship outlined above becomes more overt. This is because policy makers setting up a system pro-actively to obtain the human tissues used in SC research need to determine whether and, if so, how their providers should be entitled to payment for this provision. A consideration of the factors which lead embryo progenitors to agree to make spare embryos available to SC research will assist in identifying the role they see themselves as playing in the SC research arena. This can thus help determine the stance which a regulatory system should adopt in ensuring that the progenitors' autonomous wishes are respected as far as possible, and in establishing a system which will provide adequate and appropriate recognition of this contribution. To this end, a number of empirical studies which have explored such views will be analysed. These studies do not include information about the moral status of the embryo and carry no normative relevance. However, in setting out the providers' views about the basis on which they contribute their tissues to such research, they can be informative about whether or not they correlate with the concept of property in tissues.

In a UK-based study, 54% of 300 couples undergoing IVF consented to make their surplus embryos available for research, having been provided with information about the research by a practitioner not involved in their treatment at the scan two days before ova retrieval.[41] Those couples who consented to provide embryos for research had a significantly higher number of follicles at the scan two days before retrieval and of embryos retrieved,[42] suggesting that this gave them greater confidence in the success of their own treatment despite providing some embryos to research. Couples who had previously failed to conceive with fertilisation treatment were less likely to consent. 94% of couples who decided to donate did so on the day of oocyte retrieval, "indicating a decision made regardless of the fertilization outcome".[43]

A variety of reasons were given by those not consenting to provision to research, including conflicts between the couple about the decision and the fact that there was no limit to or control over the duration for storage and use of SCs.[44] The fact that NHS funds were not available to cryopreserve embryos may have been a factor in some couples deciding to donate.[45] Nevertheless, this study indicates that where IVF patients are fairly confident that their own reproductive aims will be met, they are willing to support the research endeavour by providing their spare embryos for research.

In a subsequent study, Hammarberg *et al.* conducted an anonymous postal survey of couples who had contacted a given clinic to make a decision about embryos they had in storage. The most common decision made was to donate the embryos to research (42%), something which the authors conjecture may be due to a growing public awareness of the issues arising out of such decisions.[46] 51% of those choosing disposal said that they did not want a full sibling of their own child to be born into another family, while 43% did not want research performed on them; of those who donated for research, 92% wanted to help advance science while 65% did not want the embryos to be wasted; and of those who chose to donate to another couple, 100% wanted to help another infertile couple and 70% wanted to give the embryos a chance at life.[47] 69% said they would have considered donating to SC research if it had been possible, citing as reasons wanting to help advance science, improve others' quality of life and not wanting to waste the embryos.[48] Those who would not have considered this gave as reasons that they considered embryos as early life which should not be used in research and wanted to allow them a chance at life with another couple.[49]

Other researchers have asked embryo providers in the UK and Switzerland to provide their views on the status of their embryos which had been created for IVF purposes.[50] In the UK cohort, couples undergoing IVF treatment were asked to consider making their spare, fresh embryos available to SC research, and the closeness in time and space to their own treatment influenced their discussion of the embryos being framed in 'baby talk'.[51] Nevertheless, of the 44 couples interviewed, most had consented to make embryos available for research, and some recognised a difference in status between those that were implanted and those which were considered unsuitable for treatment purposes.[52] The embryos' value for research purposes was acknowledged by some,[53] but the majority viewed their embryos as a precious reproductive resource, with its value as a research raw material being "very much a secondary consideration".[54]

In contrast, the Swiss couples who were approached with the request to provide their embryos for research were considering the fate of frozen embryos, often some years after the commencement of IVF treatment. Of those who decided to accede to this request, the majority did so on the basis that they wished their embryo to be put to good use. In this context, "donation gives the embryo meaningful social status as a contribution to a worthwhile endeavour, rather than being a meaningless piece of waste".[55] This study's authors were of the view that the differing approaches of couples to their embryos were context dependent. They argue that "embryos are not fixed, universal biological entities but are defined by, and acted upon in relation to, their social context, that is, by their location in time and space": this location influencing the view of the couple of the status of the embryo at a given time.[56] Influences on this context included the stage the couple had reached in their reproductive attempts, whether the embryos were fresh or frozen, the current and anticipated state of science and in the context of the legislative framework of the relevant jurisdiction.[57] "Stem cell science adds to this diversity by increasing awareness of the fluidity of the status of the embryo by introducing further uses for it, and by reinforcing the view that it is a particularly precious resource".[58]

In a Belgian study, the 'embryo disposition decisions' of seven couples and 11 women receiving fertility treatment were explored.[59] Belgian law dictates that such a decision must be made prior to treatment beginning.[60] A two-stage approach to decision-making was observed. In the first stage, themes of a genetic link to the embryo progenitors and of symbolic meaning (particularly the embryo as a symbol of the relationship between the couple) were prevalent and led to a reluctance to donate to others for reproductive purposes.[61] Once this decision had been made, "they spontaneously went to stage two of the decision sequence, in which they considered whether or not to donate for science".[62] As a result of a "lack of confidence in medical science", four patients preferred to have their embryos destroyed than to donate it to science.[63] The remainder viewed the embryo as having a "high instrumental value"(for both reproductive and research purposes),[64] illustrating embryo progenitors' awareness of the multiple forms of value possessed by the embryo.

The information from these studies gives us an insight into the reasons underlying decisions about the ultimate fate of IVF embryos and allows us to identify the views which progenitors have about the nature of the embryos they have decided to make available to research. They indicate that individuals and couples will have a wide variety of reasons for reaching their decision about the fate of their supernumerary embryos. Once the progenitors have made the decision to make their embryos available to research, on the basis of these studies it is entirely appropriate to view embryos as a research resource, thus disregarding their potential within a reproduction framework. The regulatory regime respects the autonomous wishes of these individuals in making that decision and should allow for recompense to be made for that material on the basis that it is a form of property. On reaching the decision to make their embryo available for research, the gamete providers become property owners who may offer this product to researchers for commercial benefit (or, if they choose, as a true gift). Legal recognition of their property relationship with the embryo created from their gametes forms a fundamental part of respecting their autonomous wishes in regard to that entity where they wish to make it available for research purposes.

2.3 Recompense to Providers of 'Spare' IVF Embryos

The preceding sections have attempted to establish the grounds for the conclusion that embryos made available to SC research are property. On the basis of this, embryo progenitors have the 'right to the capital', that is, the power to alienate the property and transfer title to another by way of sale.[m,65] The question of the level and type of recompense which should be made available therefore arises. It is suggested that the most practical and pragmatic manner of providing recompense would be to make a tariff-based payment for the provision of the gametes used to create the embryo rather than for the embryo itself. This permits adequate acknowledgement of the distinct roles provided by each gamete provider. It means that providers who are not in a relationship with each other do not become financially entangled as a result of the recompense provisions. Thus, for example, the gamete providers for those embryos which were formed for reproductive purposes through the use of anonymous donor gametes would be paid for directing their cells for research use once they were no longer to be used in a reproductive setting.

The tariff approach lends itself to comparisons with other situations in which a financial value is placed on uses of the body or its parts, such as in awards of compensation for personal injury or payment for participation in clinical research. In the former context, guideline amounts for injury to each part of the body have been set out,[66] and a final figure is arrived at by taking into account the specific circumstances of the individual concerned and the effect of the injury upon them. In the latter, payments to research volunteers can incorporate elements reflecting "time, inconvenience, travel or incurring risk", although "payments should not be so high as to induce people to incur a risk which is perceived as high [or there is] a low probability of a serious adverse event".[67] Sperm providers cannot be said to be undergoing physical harm or the risk of it in their contribution to the creation of the embryo and therefore an analogy with the position of a research subject, i.e. an individual who undergoes physical risk as part of their participation in the testing of a proposed medicine or technique, or

[m] Note that such transfer can take place for no financial gain if that is the wish of the tissue provider.

a victim of personal injury, cannot be made. He should be entitled to a nominal payment under the tariff system to recompense him for the time and inconvenience caused in providing his gametes and as an acknowledgement of his contribution to efforts to further SC science.

In contrast, in undergoing IVF, female progenitors are exposed to the physical discomfort and risks entailed in stimulating superovulation and in ova retrieval.[n,68,69] Even though these arise due to them consenting to undergo this procedure for their own reproductive purposes, the purpose for which the woman exposed herself to these risks and interventions is immaterial to the question of whether or not she is entitled to financial recompense for this contribution. The fact is that without her having done so, the potential asset to SC research, her embryo, would not be available. It is therefore legitimate to take into account her endeavours, and the physical effects of them upon her, in the same way that we might in considering recompense for a research subject or in calculating damages in a personal injury claim. As a result, an argument can be made that at least in terms of the physical risk to which the woman has been exposed, she would be entitled to greater recompense than her male counterpart, and it is suggested that a payment of around £2,500 would be reasonable.[70]

3. PROTECTING THE INTERESTS OF THE EMBRYO PROVIDERS WITHIN A PROPERTY FRAMEWORK

We have seen that, whatever the multitude of views which exist about the moral status of embryos, the current legal system permits practices through which more embryos are often created than will be used in the attempt to reproduce and will thus, either through disposal or research, be destroyed. It is the views of the progenitors themselves which determine which of these two options will be used. Given this central role in the process, it is important that their interests are protected within this context. Three potential threats to their interests will be considered here. The first is the potential vulnerability of

[n] See notes 68 and 69 for an account of these risks.

embryo providers when faced with requests to provide embryos for research with their therapeutic relationship with clinician-researchers. The second is the ethical objection that allowing progenitors to choose to provide embryos for research within a system which provides recompense commodifies a human entity. The final issue is the practical question of how the regulatory system should respond where the two individuals whose gametes have been combined to create an embryo have differing views, or are ambivalent, about what its fate should be.

3.1 Potential Conflicts between Patient and Tissue Provider Roles

There have been calls by SC scientists for changes in IVF practices to ensure that those embryos which become available to them are in a state which is more likely to facilitate the harvesting of SCs. For example, "a shift in practice from the cryopreservation of all embryo stages to that of only blastocysts"[71] may allow the retrieval of more SCs due to the fact that scientists have more success in creating embryonic SC lines from entities frozen at blastocyst stage than at the earlier cleavage-stage embryos. In IVF treatment, allowing development to blastocyst stage can "improve the synchronicity of uterine and embryonic development and provide a mechanism for self-selection of viable embryos".[72] Studies have shown that this approach does not affect the miscarriage, clinical pregnancy or live birth rates per cycle, however:

> [B]lastocyst transfer was associated with an increase in failure to transfer any embryos in a cycle and a decrease in embryo freezing rates. In the absence of data on cumulative live birth rates resulting from fresh and thawed cycles, it is not possible to determine if this represents an advantage or disadvantage.[73]

More information is therefore required to determine whether a policy decision to freeze at the blastocyst stage would result in a less effective level of fertility treatment for patients. Given that the embryos are usually being put into storage so that such future reproductive treatment remains an option, the provision of less than optimal quality treatment by clinicians would conflict with the legal and ethical duties of care owed to these patients. Mechanisms need to be implemented in order to ensure that patients undergoing fertility treatment are clear that requests made by their clinician for provision of their embryos to research do not represent advice that this is in their best reproductive interests.[o,74]

This is particularly relevant where any fresh embryos are deemed of sufficient quality for reproductive purposes but are requested to be provided to SC research. In such cases, a conflict exists between the couple's interest in attempting to reproduce and in science's desire to make progress in SC research.[75] There is some dispute in the scientific literature about whether frozen or fresh embryos are most suitable for the requirements of SC research,[76] meaning that evidence which would provide a justification for direct requests for fresh embryos is lacking. Policy makers, therefore, need to ensure that any requests to women or couples to donate embryos to SC research do not unnecessarily and unreasonably impair the chances of realising such aims. This could involve a stipulation that only frozen embryos be made available on the basis that the decision to provide embryos for research could be made after the outcome of reproductive attempts are known, thus avoiding any conflict of the reproductive aim and the desire to contribute to research.[77] In the alternative, such a stance might represent unnecessary paternalism. The regulatory system should ensure that the scientific methods used are those which have optimal chances of achieving the reproductive aims of embryo progenitors. Where those aims are no longer pursued, it should appropriately acknowledge their new role as raw materials providers for research, where that is the option they choose, continuing to ensure that appropriate levels of counselling and information are provided to support embryo providers in their decisions.

3.2 Commodification

Opponents of the contention that embryos can be classified as research raw materials and thus property once their progenitors have decided they

[o] See notes 33 and 34 for the provisions in force to separate clinical and research roles.

should be used for this purpose may counter that such a position commodifies the human body and is contrary to human dignity.[78] Brownsword observes a "new turn in the rhetoric of bioethics towards a 'dignitarian alliance' whose fundamental commitment is to the principle that human dignity should not be compromised".[79] This view asserts that even in the face of informed consent by participants in research, human dignity (particularly of entities not protected by human rights regulation, such as embryos), can be compromised.[80] To treat a thing solely as a commodity is to treat it as though it has a price in the marketplace; it is fungible (i.e. interchangeable with other like goods) and has only instrumental, rather than intrinsic, value.[81] Wrongful commodification has been said to take place where objects are treated as commodities despite their "not being (really) fungible or merely instrumentally valuable".[82] To treat persons as merely of instrumental value breaches the Kantian principle of not using another person merely as a means to an end, rather as an end in themselves.[p,83] It has been asserted that "conceptualizing aspects of one's person as alienable property commensurable with other properties on the market erodes the very notion of what it is to have a self".[84]

However, it seems illogical to argue that commodification of the embryo within a research context is wrong. Those who persist in pursuing this argument must necessarily object to any party, whether tissue providers, biotechnology companies, scientists or a regulator obtaining commercial benefit from them. To deny that any form of property exists in these tissues (in addition to the existing denial of these rights to their progenitors) belies the realities of the legal system governing human tissue exchange: This allows that once provided for research, embryos can be used in the attempt to achieve both commercial gains and the aim of improving levels of health for humankind. This practice of 'downstream commodification' allows for a variety of categories of individuals and organisations to make financial gains from this provision.[85] It is unfair to deny providers the opportunity to share in those potential gains entirely.

---

[p] "Act so you treat humanity, whether in your person or that of another, always as an end and never as a means only." See note 83.

To deny property rights in all forms of human tissue, including, for example, blood or skin cells, on the basis of an affront to human dignity is an over statement of the dangers of commercialism in the SC research sphere as "this interpretation requires equating the moral worth of [such] cells to that of human beings".[86] Such an approach would pose the danger of stultifying the research endeavour as restrictions were placed on the uses to which such tissue could be put as a result of an unrealistic assessment of their moral worth. Property in some types of tissue, such as hair and sperm, is already recognised in law.[q,87,88] It is hard to justify differentiating between these and other forms of bodily tissue, particularly in light of the potential benefit to society and individuals within it which might accrue from tissues used in SC research. This concept pervades the biotechnology industry at every level beyond that of tissue providers. Rather than establishing an over-protection of tissues on the basis of an exaggerated assessment of their moral worth, appropriate regulation of the SC arena must ensure that adequate measures to protect the interests of their providers, such as providing them with appropriate recompense, are in place. The dangers of commodification are less serious than an undermining of the availability of resources for SC research.

It might be argued that the issue of human dignity is not as easily dismissible in relation to embryos as it is in relation to other forms of tissues, for example regenerative tissues such as skin and blood, on the grounds of their potential to become born human beings with significant moral value and deserving of protection of their dignity. Permitting embryos to be classified as property and, thus, able to be bought and sold has been said to be "closely analogous to the selling of babies",[89] a clear infringement of the moral status of a human being. However, the commodification argument in relation to embryos used for research is misplaced. If concerns are based on the moral status of the embryo, objections should be targeted at their use and destruction in this endeavour (and even the creation of supernumerary embryos within IVF leading to

[q] See for example *R v Herbert* (note 87), in which hair was characterised as property when a conviction for larceny was imposed on a defendant who cut hair clippings from a female passenger in his car and see *Yearworth* (note 88).

their use in research or their destruction), rather than the fact that they assume a commercial value within the research sphere. The effect of such a stance would be to stifle advances in SC research,[r] given that even the development of techniques using adult cells relies on comparisons with results with embryonic SCs as the gold standard.

It is not denied that where individuals have made a decision to put an embryo to reproductive use, both the wishes of the progenitors and the embryo as an entity in itself are deserving of respect. However, where the progenitors have decided to make such embryos available for research, thus denying the embryo the chance of ever being born, the decision to use it in what is hopefully beneficial research is compatible with its new legal status as a raw material for therapeutic research. While it may seem arbitrary to leave the determination of an embryo's fate to the whim of its progenitors, a consensus on embryonic moral status is far from being achieved. Thus, where an embryo can be put to good use, it does not seem unreasonable to allow those who have contributed their gametes to its creation to determine its legal status and use.[s,90]

Viewing embryos as raw materials in the research context does not exclude the possibility of valuing them in different ways in different contexts. On this basis one can treat human beings as entities with context-dependent values, dictated by the progenitors within the limits prescribed by law, without violating their dignity. Rather than an outright exclusion of commercial influences in biotechnological research, we should consider whether the appropriate level, form and extent of property which is permitted in tissues has been arrived at, and whether appropriate recognition of any other values is available under the regulatory framework. Such values can be adequately protected through

[r] Not to mention implications for the health and autonomy of women receiving assisted conception treatment where entities with equivalent moral value were being created through that process, which might lead for example to a requirement all embryos created be placed in their body.

[s] An argument for regulating the use of embryos according to whether they are put to reproductive or research purposes has also been made by Johnson (note 90).

regulatory mechanisms without the need to deny their progenitors just recompense for their contribution to science.

3.3 Disagreement or Disengagement on the Fate of the Embryo

We have seen that the law allows embryo progenitors to determine the ultimate fate of their embryos. However, the fact that the provision of embryonic tissue to SC research depends on the two providers agreeing on the appropriate course of action, and the fact that considerable time may elapse between the providers receiving treatment and having to finalise this decision, mean that complexities can arise in obtaining clear indications of the progenitors' decision. Embryos might be 'abandoned' as patients decline to respond to a clinic's request to decide whether they should be destroyed or used in research projects, or the couple may disagree on the appropriate course of action.

The Joint Committee on the Human Tissues and Embryos (Draft) Bill recommended that couples undergoing IVF should be asked to consent to any embryos or gametes still in storage at the end of the statutory storage period becoming the "property of the HFEA"and then being used for research purposes.[91] Such consent could be withdrawn at any time during the storage period.[92] However, this position was rejected by the government[93] because it would mean that patients would have to decide, before commencing treatment, whether to consent to making spare embryos available for research. It preferred instead for patients' treatment decisions to be concluded before determining the fate of 'spare' embryos in light of knowledge of specific research for which they are intended.[94] The resulting default position is that where no decision is made, this potential raw material for SC research will be destroyed, a position which shows a disregard for a value of any kind for these entities.

The only circumstance in which the law provides for any advance and definitive indication about the embryo providers' decision is the requirement that individuals consenting to the storage of embryos or gametes must state in writing what should happen to those tissues should the individual die or become mentally incapacitated.[95] Such a decision should only be made after the provision of necessary information and appropriate counselling, and can be varied or withdrawn at any time up to the use of the embryo or gametes in treatment or research.

It is submitted that UK policy makers should adopt the system now used in Belgium in which advance directives (ADs) dictating what should happen to embryos created through IVF are made as a matter of routine and cover a variety of situations in which the living, capacitous progenitor may find themselves.[96,97] Such documents should be drawn up at the time when the decision is made to store embryos. Going beyond the limited circumstances of death and incapacity which the 1990 Act and HFEA Code of Practice anticipate, they should clearly state the circumstances in which they will come into effect, such as death, loss of capacity, divorce, untraceability, disagreement, or failure to pay storage fees.[98] Where the provider remains prepared to give information to the clinic about their wishes, the AD will not be relevant or legally enforceable. The legal default position for the embryos of those refusing to provide an indication should be provision to research, rather than destruction. Appropriate counselling, to the same standard and extent as that for obtaining consent to treatment and storage must be given[99] in order for the provider to understand the consequences of not being able or willing to make a decision about what should happen to their embryos on expiry of the storage period, and in order that they are aware of the potential uses to which the embryos will be put, including SC research.[100] The AD provider should be informed about any changes in the law, treatment success rates, clinic policies etc so that they can decide on an on-going basis whether these affect their wishes.[101]

It may be objected that the suggested change to the default position is unethical. However, social utility arguments can be made that, where progenitors do not wish to continue to use their embryos for reproductive aims, society would benefit from them being put to beneficial research uses. While the suggestion is not being made here that all spare embryos should be put to such use, where no objections have been made and where the progenitors have been made aware that this will be the outcome, it seems ethical to proceed in this manner. This can be compared to the historical legal doctrine of abandonment which takes place when possession of goods is quitted voluntarily without any intention of transferring them to another.[102] The finder of a chattel acquires no rights over it unless it has been abandoned or lost and he takes it into his care and control.[103] Under these circumstances, no one would suggest that the abandoned property should automatically be destroyed. In the approach suggested here, the person who abandons an embryo is given a personal notification of the consequences of such an action, meaning that it is entirely reasonable to use the embryo as stated. Where they are willing to give an indication of their wishes, as the owners of the embryos, those wishes must be respected.

4. CONCLUSION

This chapter has focussed on regulatory issues arising in relation to the use of spare IVF embryos in SC research. In the first part, the argument that tissues provided for SC research should be considered the property of its providers was tested in relation to such embryos, leading to the conclusion that they are property in the research context. It was argued that while embryos can have certain values in a reproductive setting, once the individuals whose gametes have been combined to form the embryo have decided that they should be made available for research purposes, the law allows for them to assume the status of property, for which appropriate recompense should be made. The fact that the current regulatory system fails to acknowledge this contribution is something for which it can rightly be criticised and which should be remedied through the implementation of a regime which incorporates appropriate financial recognition.

The final part of this chapter attempted to allay concerns which might be raised by the preceding conclusion. None of these concerns constituted compelling reasons to deny property status to embryos used in SC research, but served to confirm that appropriate regulation which is fair to all participants in the SC research endeavour may contribute to the realisation of the hopes for SC technology.

REFERENCES

1. *Yearworth and Others v North Bristol NHS Trust* [2009] EWCA Civ 37.
2. Annas GJ. (1999) Waste and Longing — the Legal Status of Placental-Blood Banking. *Legal Issues in Medicine* **340(19):** 1521–1524, p. 1521.
3. Takahashi K, *et al.* (2007) Induction of Pluripotent Stem Cells form Adult Human Fibroblasts by Defined Factors. *Cell* **131:** 861–72; Yu J, *et al.* (2007) Induced Pluripotent Stem Cell Lines Derived From Human Somatic Cells. *Science* **318:** 1917–1920.
4. Lerou PH, Yabuuchi A, Huo H *et al.* (2008) Human Embryonic Stem Cell Derivation from Poor-quality Embryos. *Nature Biotechnology* **26:** 212–214.
5. Barratt CLR, St John JC, Afnan M. (2004) Clinical Challenges in Providing Embryos for Stem-Cell Initiatives. *The Lancet* **364:** 115–118, p. 115.
6. Barratt CLR, St John JC, Afnan M. (2004) Clinical Challenges in Providing Embryos for Stem-Cell Initiatives. *The Lancet* **364:** 115–118, p. 117.
7. Human Fertilisation and Embryology Authority, Code of Practice (8th Edition), section 7.
8. Human Fertilisation & Embryology Authority (October 2006) *One Child at a Time.* Available at http://www.hfea.gov.uk/docs/One_at_a_time_report.pdf (Accessed 16 June 2011).
9. Human Fertilisation & Embryology Authority (October 2007) *Multiple Births and Single Embryo Transfer Review: Evidence Base and Policy Analysi*s (HFEA (17/10/07) 401). Available at http://www.hfea.gov.uk/docs/AM_MB_and_SET_review_Oct07.pdf (Accessed 16 June 2011).
10. Stacey G, Hunt CJ. (2006) The UK Stem Cell Bank: A UK Government-Funded, International Resource Center for Stem Cell Research. *Regenerative Medicine* **1:** 139–142.
11. UK Stem Cell Bank. Stem Cell Catalogue. Available at http://www.uk stemcellbank.org.uk/stemcelllines/stemcellcatalogue.cfm (Accessed 16 June 2011).
12. See for example Resnik DB. (2002) The Commercialization of Human Stem Cells: Ethical and Policy Issues. *Health Care Analysis* **10(2):** 147.
13. Franklin S. (2006) Embryonic Economies: The Double Reproductive Value of Stem Cells. *Biosocieties* **1(1):** 71–90, p. 72.
14. Human Fertilisation and Embryology Authority, Code of Practice (8th Edition), section 7.
15. Human Fertilisation and Embryology Act 1990 (as amended 2008), section 14(4).

16. Human Fertilisation and Embryology Act 1990 (as amended 2008), section 14(1)(c).
17. Honoré AM. (1987) *Making Law Bind: Essays Legal and Philosophical.* Clarendon Press, Oxford.
18. Honoré AM. (1987) *Making Law Bind: Essays Legal and Philosophical.* Clarendon Press, Oxford, p. 165.
19. Human Fertilisation and Embryology Act 1990, paragraph 6(3), schedule 3.
20. Human Fertilisation and Embryology Authority, Code of Practice (8th Edition), section 22B.
21. Human Fertilisation and Embryology Act 1990, schedule 3, paragraph 1(3).
22. Human Fertilisation and Embryology Act 1990, paragraphs 6(1) and (3).
23. Knowles LP. (1999) Human Primordial Stem Cells: Property, Progeny and Patents. *The Hastings Center Report* **29(2):** 38.
24. Human Fertilisation and Embryology Act 1990 (as amended 2008), schedule 3, paragraph 4A.
25. Human Fertilisation and Embryology Authority, Code of Practice (8th Edition), section 22.7(j).
26. Honoré AM. (1987) *Making Law Bind: Essays Legal and Philosophical.* Clarendon Press, Oxford, pp. 168–169.
27. Human Fertilisation and Embryology Act 1990, schedule 3, paragraph 1.
28. Human Fertilisation and Embryology Act 1990, paragraph 2(1).
29. Human Fertilisation and Embryology Act 1990, paragraph 3.
30. Human Fertilisation and Embryology Authority, Code of Practice (8th Edition), R19 and 22.6. See paragraph 22.7 for further information required to be given.
31. Human Fertilisation and Embryology Authority, Code of Practice (8th Edition), section 22.
32. Heng BC. (2006) Donation of Surplus Frozen Embryos for Stem Cell Research or Fertility Treatment — Should Medical Professionals and Healthcare Institutions Be Allowed to Exercise Undue Influence on the Informed Decision of Their Former Patients? *Journal of Assisted Reproduction and Genetics* **23:** 381–382.
33. Human Fertilisation and Embryology Authority Code of Practice, (8th Edition), paragraph 22.13.
34. Human Fertilisation and Embryology Authority Code of Practice, (8th Edition), paragraph 22.14.
35. Grubb A. (1998) 'I, Me Mine': Bodies, Parts and Property. *Medical Law International* **3:** 299–317.

36. Grubb A. (1998) 'I, Me Mine': Bodies, Parts and Property. *Medical Law International* **3:** 299–317, p. 166.
37. *Yearworth and Others v North Bristol NHS Trust* [2009] EWCA Civ 37.
38. *Yearworth and Others v North Bristol NHS Trust* [2009] EWCA Civ 37. [45(f)(ii)].
39. Human Fertilisation and Embryology Act 1990, s12(e).
40. Devolder K, Harris J. (2007) The Ambiguity of the Embryo: Ethical Inconsistency in the Human Embryonic Stem Cell Debate. *Metaphilosophy* **38:** 153–169, p. 153.
41. Choudary M, Haimes E, Herbert M, *et al.* (2004) Demographic, Medical and Treatment Characteristics Associated With Couples' Decisions to Donate Fresh Spare Embryos for Research. *Human Reproduction* **19(9):** 2091–2096.
42. Choudary M, Haimes E, Herbert M, *et al.* (2004) Demographic, Medical and Treatment Characteristics Associated With Couples' Decisions to Donate Fresh Spare Embryos for Research. *Human Reproduction* **19(9):** 2091–2096.
43. Choudary M, Haimes E, Herbert M, *et al.* (2004) Demographic, Medical and Treatment Characteristics Associated With Couples' Decisions to Donate Fresh Spare Embryos for Research. *Human Reproduction* **19(9):** 2091–2096, p. 2092.
44. Choudary M, Haimes E, Herbert M, *et al.* (2004) Demographic, Medical and Treatment Characteristics Associated With Couples' Decisions to Donate Fresh Spare Embryos for Research. *Human Reproduction* **19(9):** 2091–2096, p. 2093.
45. Choudary M, Haimes E, Herbert M, *et al.* (2004) Demographic, Medical and Treatment Characteristics Associated With Couples' Decisions to Donate Fresh Spare Embryos for Research. *Human Reproduction* **19(9):** 2091–2096, p. 2095.
46. Hammarberg K, Tinney L. (2006) Deciding the Fate of Supernumerary Frozen Embryos: A Survey of Couples' Decisions and the Factors Influencing Their Choice. *Fertility and Sterility* **86(1):** 86–91, pp. 86–87 & 89.
47. Hammarberg K, Tinney L. (2006) Deciding the Fate of Supernumerary Frozen Embryos: A Survey of Couples' Decisions and the Factors Influencing Their Choice. *Fertility and Sterility* **86(1):** 86–91, p. 88.
48. Hammarberg K, Tinney L. (2006) Deciding the Fate of Supernumerary Frozen Embryos: A Survey of Couples' Decisions and the Factors Influencing Their Choice. *Fertility and Sterility* **86(1):** 86–91, p. 89.
49. Hammarberg K, Tinney L. (2006) Deciding the Fate of Supernumerary Frozen Embryos: A Survey of Couples' Decisions and the Factors Influencing Their Choice. *Fertility and Sterility* **86(1):** 86–91, p. 89.

50. Haimes E, Porz R, Scully J, Rehmann-Sutter C. (2008) "So What is an Embryo?" A Comparative Study of the Views of Those Asked to Donate Embryos for hESC Research in the UK and Switzerland. *New Genetics and Society* **27(2):** 113–126.
51. Haimes E, Porz R, Scully J, Rehmann-Sutter C. (2008) "So What is an Embryo?" A Comparative Study of the Views of Those Asked to Donate Embryos for hESC Research in the UK and Switzerland. *New Genetics and Society* **27(2):** 113–126.
52. Haimes E, Porz R, Scully J, Rehmann-Sutter C. (2008) "So What is an Embryo?" A Comparative Study of the Views of Those Asked to Donate Embryos for hESC Research in the UK and Switzerland. *New Genetics and Society* **27(2):** 113–126, p. 116.
53. Haimes E, Porz R, Scully J, Rehmann-Sutter C. (2008) "So What is an Embryo?" A Comparative Study of the Views of Those Asked to Donate Embryos for hESC Research in the UK and Switzerland. *New Genetics and Society* **27(2):** 113–126, p. 118.
54. Haimes E, Porz R, Scully J, Rehmann-Sutter C. (2008) "So What is an Embryo?" A Comparative Study of the Views of Those Asked to Donate Embryos for hESC Research in the UK and Switzerland. *New Genetics and Society* **27(2):** 113–126, p. 119.
55. Haimes E, Porz R, Scully J, Rehmann-Sutter C. (2008) "So What is an Embryo?" A Comparative Study of the Views of Those Asked to Donate Embryos for hESC Research in the UK and Switzerland. *New Genetics and Society* **27(2):** 113–126, p. 122.
56. Haimes E, Porz R, Scully J, Rehmann-Sutter C. (2008) "So What is an Embryo?" A Comparative Study of the Views of Those Asked to Donate Embryos for hESC Research in the UK and Switzerland. *New Genetics and Society* **27(2):** 113–126, pp. 124–125.
57. Haimes E, Porz R, Scully J, Rehmann-Sutter C. (2008) "So What is an Embryo?" A Comparative Study of the Views of Those Asked to Donate Embryos for hESC Research in the UK and Switzerland. *New Genetics and Society* **27(2):** 113–126, p. 125.
58. Haimes E, Porz R, Scully J, Rehmann-Sutter C. (2008) "So What is an Embryo?" A Comparative Study of the Views of Those Asked to Donate Embryos for hESC Research in the UK and Switzerland. *New Genetics and Society* **27(2):** 113–126, p. 125.
59. Provoost V, *et al.* (2009) Infertility Patients' Beliefs about their Embryos and their Disposition Preferences. *Human Reproduction* **1(1):** 1–10.

60. Provoost V, *et al.* (2009) Infertility Patients' Beliefs about their Embryos and their Disposition Preferences. *Human Reproduction* **1(1):** 1–10, p. 2.
61. Provoost V, *et al.* (2009) Infertility Patients' Beliefs about their Embryos and their Disposition Preferences. *Human Reproduction* **1(1):** 1–10, pp. 8–9.
62. Provoost V, *et al.* (2009) Infertility Patients' Beliefs about their Embryos and their Disposition Preferences. *Human Reproduction* **1(1):** 1–10, p. 6.
63. Provoost V, *et al.* (2009) Infertility Patients' Beliefs about their Embryos and their Disposition Preferences. *Human Reproduction* **1(1):** 1–10, p. 6.
64. Provoost V, *et al.* (2009) Infertility Patients' Beliefs about their Embryos and their Disposition Preferences. *Human Reproduction* **1(1):** 1–10, p. 6.
65. Honoré AM. (1987) *Making Law Bind: Essays Legal and Philosophical.* Clarendon Press, Oxford, p. 170.
66. Judicial Studies Board. (2008) *Guidelines for the Assessment of General Damages in Personal Injury Cases,* 9th Edition. Oxford University Press, Oxford.
67. Royal College of Physicians. (2007) Guidelines on the Practice of Ethics Committees in Medical Research with Human Participants, 4th Edition [10.13].
68. Balen A. (February 2005) *Ovarian Hyperstimulation Syndrome — A Short Report for the HFEA.* http://www.hfea.gov.uk/cps/rde/xbcr/hfea/OHSS_Report_from_Adam_Balen_2005(1).pdf (Accessed 16 June 2011).
69. Balen A. (August 2008) *Ovarian Hyperstimulation Syndrome — A Short Report for the HFEA.* http://www.hfea.gov.uk/docs/OHSS_UPDATED_Report_from_Adam_Balen_2008.pdf (Accessed 16 June 2011).
70. A more detailed account of the reasons for settling upon this amount are provided in Devaney S. (2010) Tissue Providers for Stem Cell Research: The Dispossessed. *Law, Innovation and Technology* **2(2):** 165–191.
71. Barratt CLR, St John JC, Afnan M. (2004) Clinical Challenges in Providing Embryos for Stem-Cell Initiatives. *The Lancet* **364:** 115–118, p. 115.
72. Blake D, Proctor M, Johnson N, Olive D. (2004) Cleavage Stage versus Blastocyst Stage Embryo Transfer in Assisted Conception: A Cochrane Review. *Human Reproduction* **19(9):** 2174.
73. Blake D, Proctor M, Johnson N, Olive D. (2004) Cleavage stage versus blastocyst stage embryo transfer in assisted conception: A Cochrane Review. *Human Reproduction* **19(9):** 2174.
74. McLeod C, Baylis F. (2007) Donating Fresh versus Frozen Embryos to Stem Cell Research: In Whose Interests? *Bioethics* **21(9):** 465–477.
75. McLeod C, Baylis F. (2007) Donating Fresh versus Frozen Embryos to Stem Cell Research: In Whose Interests? *Bioethics* **21(9):** 465–477.

76. McLeod C, Baylis F. (2007) Donating Fresh versus Frozen Embryos to Stem Cell Research: In Whose Interests? *Bioethics* **21(9):** 465–477.
77. As recommended by the Ethics Committee of the American Society for Reproductive Medicine. (2002) Donating Spare Embryos for Embryonic Stem Cell Research. *Fertility and Sterility* **78(5):** 957–960.
78. Lebacqz K. (2001) Who 'Owns' Cells and Tissues? *Health Care Analysis* **9:** 353–367, p. 355.
79. Brownsword R. (2004) Regulating Human Genetics: New Dilemmas for a New Millennium. *Medical Law Review* **12(1):** 14–39, p. 20.
80. Brownsword R. (2004) Regulating Human Genetics: New Dilemmas for a New Millennium. *Medical Law Review* **12(1):** 14–39, p. 20.
81. Wilkinson S. (2007) Commodification. In: Ashcroft RE, Dawson A, Draper H, McMillan JR (eds.), *Principles of Health Care Ethics.* John Wiley and Sons, Chichester, pp. 285–291.
82. Wilkinson S. (2007) Commodification. In: Ashcroft RE, Dawson A, Draper H, McMillan JR (eds.), *Principles of Health Care Ethics.* John Wiley and Sons, Chichester, pp. 285–291, p. 286.
83. Kant I. (1785) *Groundwork of the Metaphysics of Morals.* White Beck L (trans). *Foundations of the Metaphysics of Morals* (Library of Liberal Art: 1959 (VI)), 429.
84. Holland S. (2001) Contested Commodities at Both Ends of Life: Buying and Selling Gametes, Embryos, and Body Tissues. *Kennedy Institute of Ethics Journal* **11(3):** 263–284, p. 271.
85. Holland S. (2001) Contested Commodities at Both Ends of Life: Buying and Selling Gametes, Embryos, and Body Tissues. *Kennedy Institute of Ethics Journal* **11(3):** 263–284, p. 266.
86. Korobkin R. (2007) Buying and Selling Human Tissues for Stem Cell Research. *Arizona Law Review* **49:** 45–67, p. 56.
87. *R v Herbert* [1961] JPLGR 12, 13.
88. *Yearworth and Others v North Bristol NHS Trust* [2009] EWCA Civ 37.
89. Korobkin R. (2007) Buying and Selling Human Tissues for Stem Cell Research. *Arizona Law Review* **49:** 45–67, p. 57.
90. Johnson MH. (2006) Escaping the Tyranny of the Embryo? A New Approach to ART Regulation Based on UK and Australian Experiences. *Human Reproduction* **21(11):** 2756–2765, p. 2760.
91. Joint Committee on the Human Tissue and Embryos (Draft) Bill. (2007) *Report of the Joint Committee on the Human Tissues and Embryos (Draft) Bill.* The Stationery Office, London, Vol. I HL Paper 169-I, HC Paper 630-I, paragraph 250.

92. Joint Committee on the Human Tissue and Embryos (Draft) Bill. (2007) *Report of the Joint Committee on the Human Tissues and Embryos (Draft) Bill.* The Stationery Office, London, Vol. I HL Paper 169-I, HC Paper 630-I, paragraph 250.
93. Government Response to the Report from the Joint Committee on the Human Tissue and Embryos (Draft) Bill, Cm 7209, (8 October 2007), paragraph 60.
94. Government Response to the Report from the Joint Committee on the Human Tissue and Embryos (Draft) Bill, Cm 7209, (8 October 2007), paragraph 61.
95. Human Fertilisation and Embryology Act 1990, paragraph 2(2)(b) of Schedule 3.
96. Provoost V, *et al.* (2009) Infertility Patients' Beliefs about their Embryos and their Disposition Preferences. *Human Reproduction* **1(1):** 1–10.
97. Pennings G. (2000) What Are the Ownership Rights for Gametes and Embryos? *Human Reproduction* **15(5):** 979–986.
98. Pennings G. (2000) What Are the Ownership Rights for Gametes and Embryos? *Human Reproduction* **15(5):** 979–986, p. 981.
99. Pennings G. (2000) What Are the Ownership Rights for Gametes and Embryos? *Human Reproduction* **15(5):** 979–986, p. 982.
100. Ethics Committee of the American Society for Reproductive Medicine. (2002) Donating Spare Embryos for Embryonic Stem Cell Research. *Fertility and Sterility* **78(5):** 957–960, p. 959.
101. Pennings G. (2000) What Are the Ownership Rights for Gametes and Embryos? *Human Reproduction* **15(5):** 979–986, p. 982.
102. Halsbury's *Laws of England — Personal Property*, Volume 35, 5th edition, LexisNexis, London 2011 [1225].
103. *Parker v British Airways Board* [1982] QB 1004, [1982] 1 All ER 834, CA.

6

Against the Discarded–Created Distinction in Embryonic Stem Cell Research

Katrien Devolder

The discarded–created distinction is the most popular middle-ground position in the embryonic stem cell debate. It holds that it is ethically permissible to derive and use embryonic stem cells from discarded *in vitro* fertilisation (IVF) embryos but not from embryos created especially for research. I first argue that the arguments of beneficence and proportionality in support of using discarded IVF embryos show there is a presumption against the discarded–created distinction. I then consider how one might nevertheless defend the discarded–created distinction by appealing to arguments that may override this presumption. I argue that the nothing-is-lost argument fails to do this. By using discarded IVF embryos, one legitimizes and thus indirectly encourages the destruction of embryos in IVF as well as in research, so something is lost by destroying discarded IVF embryos. I further argue that the argument that the moral costs of destroying embryos in IVF are smaller than those of destroying research embryos for research does not hold either. I conclude that since there is a presumption against the discarded–created distinction, as long as no good argument has been adduced in support of it, we should reject it as a sound ethical position.

Department of Philosophy and Moral Sciences, Bioethics Institute Ghent University, Blandijnberg 2, B-9000 Ghent, Belgium. Email: Katrien.Devolder@UGent.be

1. INTRODUCTION

1.1 The Dilemma

Most of the ethical debate about whether we should support human embryonic stem cell (hESC) research turns on a fundamental disagreement about how we should treat early human embryos. As it is currently done, the isolation of human embryonic stem cells (hESCs) involves dismantling the early human embryo — a process the embryo does not survive. Many people accord significant value to the human embryo (henceforth 'embryo') and think that embryos may not simply be used in whatever way suits our research interests. However, hESC research holds unique promise for developing therapies for currently incurable diseases, as well as for important biomedical research and drug and toxicity testing. This creates a dilemma: Either one supports hESC research and accepts resulting embryo destruction, *or* one opposes hESC research and accepts that its potential benefits will be foregone.

Of course, for some people there is no dilemma. If you believe an early human embryo is merely a collection of cells that has very little value in itself, the embryo's moral status provides no reason to abstain from hESC research. On the other hand, if you believe that the embryo has the same moral status as a typical adult human, then isolating hESCs is equivalent to murder and is clearly unjustified. Yet for those who accord lesser, but still significant moral status to the embryo, the dilemma is very real.

A popular response to the dilemma has been to adopt some intermediate position between the dominant opposed ethical views that the moral status of the embryo always and never gives us decisive reasons to abstain from hESC research. One such intermediate position that has been widely defended, and serves as a basis for stem cell policy in many western countries, including the United States (US) under the Obama administration, is the so-called 'discarded–created distinction'.[1]

2. THE DISCARDED–CREATED DISTINCTION

The discarded–created distinction draws a moral line between two types of hESCs based on their origin: embryos discarded following *in vitro* fertilisation (IVF) and embryos created solely for the purpose of stem cell research.

> *Discarded–created distinction*: It is ethically permissible to derive and use hESCs from discarded IVF embryos. It is ethically impermissible to create embryos solely for the purpose of stem cell derivation, and to derive or use hESCs from such embryos.

2.1 Discarded Embryos

The great majority of currently existing hESC lines originate from embryos discarded following IVF. A woman undergoing IVF receives hormone therapy to stimulate the development and maturation of multiple eggs. After retrieval, the eggs are fertilised with semen in culture media. In most countries where IVF is practised, on average 5 to 10 embryos are produced, one or two of which will be transferred to the woman's womb to try to achieve a pregnancy. The remaining embryos are placed in nitrogen freezers, where they are cryopreserved. If an attempt to achieve a pregnancy fails, one or two embryos can be thawed for another attempt. Cryopreservation of several embryos has the advantage that women do not have to undergo the risky and uncomfortable hormone therapy and egg retrieval procedure after each failed attempt to generate a pregnancy.

At the start of their IVF treatment, individuals or couples must indicate one of the following three options for handling of any leftover embryos, that is, embryos unwanted by the individual or couple: (1) anonymous donation to infertile couples, (2) donation to scientific research, or (3) letting the embryos perish.[a] Note that the two last options both involve certain death

[a] These three options are presented in most countries where IVF is practiced. Some countries, however, only offer one or two of these options, or none. In Brazil, for example, couples can only opt for indefinite cryopreservation.

for the embryos with no opportunity to develop into a child. The great majority of existing hESC lines are obtained from leftover IVF embryos donated for research — as under option (2). I will refer to such embryos as discarded IVF embryos.

2.2 Created Embryos

There is an alternative source of hESCs. The cells could be derived from embryos especially *created* for the purpose of research. I will refer to such embryos as 'research embryos' (the term 'created' in the discarded–created distinction is somewhat confusing as IVF embryos are created too, just for a different purpose). One way to generate research embryos is through IVF using donor gametes. In some countries, stem cell scientists have encountered difficulties in achieving their intended research goals because the number of discarded IVF embryos available for stem cell research is insufficient. Creating research embryos using IVF (with no reproductive intent) could solve this problem and would allow these researchers to achieve their research goals more efficiently. But research embryos would, in the future, most likely be generated using cloning or somatic cell nuclear transfer (SCNT). SCNT involves transferring the nucleus of a somatic cell into an oocyte from which the nucleus, and thus most of the genetic material, has been removed. The manipulated oocyte is then treated, usually with an electric current, in order to stimulate cell division, and an embryo is formed. A significant advantage of using stem cells from embryos produced through SCNT rather than those derived from discarded IVF embryos is that the hESCs would be genetically identical to the person from whom the somatic cell nucleus was taken, perhaps a patient. One goal of stem cell research is the development of stem cell-based therapies. Embryonic stem cells could be used to generate an inexhaustible supply of replacement cells to regenerate damaged tissues and organs. The advantage of using hESCs that are genetically identical to the patient is that the cells or tissues produced from these stem cells would not be rejected by the patient's immune system. This is significant, as one of the major problems in transplantation medicine is immunorejection of the graft. Unfortunately, these therapies are not yet on the horizon. A great deal of research is required, including research with embryos created through SCNT; researchers will have to learn how to routinely produce such embryos and how to derive stem cells from them.

Embryonic stem cells genetically identical to the patient could also be used to study diseases *in vitro*. Researchers could create large numbers of stem cells genetically identical to the patient and then experiment on these in order to understand the particular features of the disease in that person. This increased possibility to study disease in a dish would enable research that cannot be done on patients themselves, or where there are too few patients to work with as in the case of rare genetic diseases. Stem cells genetically identical to a patient would also be of great value for drug screening and toxicity testing. For example, hepatocytes (liver cells) derived from stem cells that were obtained from embryos genetically identical to patients with various genetic and disease backgrounds could be used for predicting the liver toxicity of candidate drug therapies.

Thus, using embryos created through SCNT can help us achieve research and therapeutic aims that would be much more difficult or perhaps impossible to achieve using discarded IVF embryos. However, embryos created using SCNT are research embryos — they are created solely for the purpose of research. According to the discarded–created distinction, using (and thus destroying) research embryos is ethically unacceptable, even to achieve the aforementioned promising research and therapeutic aims. Since many countries have based their stem cell regulations on the discarded–created distinction, research using SCNT embryos has been severely restricted worldwide.

The question then is why we should we forego the benefits of research on embryos created especially for the purpose of research, be it by using SCNT or IVF? Why is the destruction of research embryos to obtain stem cells less acceptable than the destruction of discarded IVF embryos, and to such an extent that the latter should be supported but the former rejected? I start by investigating how proponents of the discarded–created distinction justify the destruction of discarded IVF embryos to obtain stem cells. I show that the justification for destroying discarded

IVF embryos also applies, at least at first sight, to the destruction of research embryos. This suggests that there should be a presumption against the discarded–created distinction. I then consider whether this presumption can be overridden. After exploring the main arguments in support of the discarded–created distinction, I conclude that, none of these arguments hold and that we should reject the discarded–created distinction as an ethical position. First, however, let us consider how proponents of the discarded–created distinction justify the destruction of discarded IVF embryos.

3. ARGUMENTS FOR USING DISCARDED IVF EMBRYOS

3.1 Beneficence

Human embryonic stem cell research could, if successful, benefit hundreds of thousands of people. It could not only save and prolong lives, but also considerably reduce morbidity. Failing to pursue this research could result in many avoidable deaths, and in more suffering in general. It is widely accepted that we have significant reasons to benefit people if we can and to prevent avoidable suffering and death. There are thus significant reasons of beneficence for pursuing hESC research. Proponents of the discarded–created distinction have appealed to these reasons to justify destroying discarded IVF embryos in stem cell research.[2,3]

3.2 Proportionality

Even if one thinks that reasons of beneficence and non-maleficence are strong reasons for pursuing hESC research, one may still think that these reasons are not strong enough to outweigh the countervailing reasons not to destroy embryos. Which reasons prevail will depend, at least in part, on how great the benefits of hESC research are relative to the moral costs of embryo destruction. Proponents of the discarded–created distinction evidently believe that the benefits are large enough, relative to the costs, to justify the destruction of discarded IVF embryos in stem cell research. This is sometimes made explicit. For example, in its recommendations on stem cell research, the US National Bioethics Advisory Commission (NBAC),[4] defending the discarded–created distinction, wrote: "In our view, the potential benefits of the research outweigh the harms to the embryos that are destroyed in the research process."[5] The obvious question that arises, then, is: If the benefits of hESC research are sufficient to outweigh the moral costs of destroying discarded IVF embryos, why do the same benefits not outweigh the costs of destroying *research* embryos? At first sight, the benefits that justify the destruction of discarded IVF embryos could also justify the destruction of research embryos. The benefits are of the same kind, and potentially, equally large. Moreover, most people would agree that the moral status of discarded IVF embryos is equal to that of research embryos. This suggests there should be a presumption in favour of treating research and discarded IVF embryos similarly — a presumption against the discarded–created distinction. I now turn to consider how one might nevertheless defend the discarded–created distinction by appealing to arguments that may override this presumption.

4. ARGUMENTS FOR THE DISCARDED–CREATED DISTINCTION

4.1 The Least Controversial Approach

An initial argument holds that, even if the benefits of hESC research outweigh both the use of discarded IVF embryos and the use of research embryos, we should only use discarded IVF embryos because, if we have different means for achieving the intended goals, we should always opt for the least controversial or the least offensive approach.[6] However, it is not true that we can achieve our intended research goals by using only discarded IVF embryos. As mentioned earlier, some research (and therapeutic) goals can only be achieved using research embryos. Furthermore, determining what approach is the least controversial is, of course, controversial itself. It will partly depend on who decides what is most controversial and on how well people are informed about the variety of issues raised by different types of stem cell research. It seems that in the stem cell debate it has been assumed too quickly that respect for the embryo is the only relevant or the most important consideration for determining how controversial certain means for obtaining stem cells are. Yet other considerations, that have garnered less attention, may also need to be taken into account; such as whether alternative means will slow down the research, what risks they involve to patients, what the economic costs will be, and so forth.[7] For example, obtaining neuronal stem cells from patients through brain biopsy does not involve embryo destruction, but may impose a significant risk to the patient and may, as a result, be more controversial than using stem cells from research embryos. Moreover, some may think that embryos created through SCNT have lower moral status than discarded IVF embryos, and that it is therefore less controversial to use SCNT embryos to obtain stem cells. It has, for example, been argued that SCNT embryos have lower moral status because they lack the potential to develop into a baby (this is not yet technically possible) or because according them moral status on the basis of their potential would result in a *reductio ad absurdum*; in McHugh's words, "then every somatic cell would deserve some protection because it has the potential to follow the same path".[8]

Stating that we should reject the use of research embryos simply because it is more controversial than using discarded IVF embryos not only assumes too quickly that both means will enable us to achieve the same research goals, but also begs the question. That using stem cells from research embryos is the most controversial approach is exactly what needs to be argued for. That leaves us with our initial question: Why is it so much worse to use research embryos than to use discarded IVF embryos?

4.2 The Nothing-Is-Lost Argument

The most important argument in defence of the discarded–created distinction is the nothing-is-lost argument (sometimes also referred to as the no-greater-loss argument).[9] This argument has been appealed to in other ethical debates. For example, Paul Ramsey appealed to it in the debate about the ethics of direct killing in situations where two lives directly collide with each other and one cannot save one without killing the other.[10] Ramsey argued that intentionally taking innocent human life is always wrong except when two conditions are met: (1) the innocent will die soon in any case, and (2) other innocent lives will be saved. In Bernard Williams' famous case 'Jim and the Indians',[11] Jim faces such a situation. Jim arrives in a South American town where 20 Indians are just about to be killed by a group of soldiers. The captain makes an offer: If Jim kills one Indian, the others will be let off. If Jim refuses the offer, the captain will do what he would do had Jim not arrived — kill the 20 Indians. Jim's killing of one Indian may be justified by the nothing-is-lost argument: The one Indian would have died soon in any case, and the other Indians will be saved.

The nothing-is-lost argument has been extended to the stem cell debate.[12–14] It has been argued that, since (1) discarded IVF embryos (innocent life) will die soon in any case and (2) patients suffering from terrible diseases and disabilities (other innocent lives) will be saved, it is permissible to destroy discarded IVF embryos to derive stem cells: It will not change their final disposition (death in the early embryonic stage) and will, thus, not cause a greater loss than would otherwise occur. It has further been argued that the nothing-is-lost argument cannot be extended to the derivation of stem cells from research embryos. According to Gene Outka, a committed defender of the discarded–created distinction, "we would distort the nothing is lost principle beyond recognition if we extended it to say that nothing is lost when we create an entity whose prospects are nil because of what we intend from the start".[15] For the nothing-is-lost argument to be applicable, it is at least required "that we encounter circumstances we did not initiate and that we wish were otherwise".[16] So it seems that for the nothing-is-lost argument to be applicable, one should not have initiated the circumstances which ensure that the loss will occur in any case, and if these circumstances obtain, one should wish they did not.

4.2.1 *Initiating Circumstances Ensuring that the Loss would Occur in Any Case*

In what sense does a researcher who derives stem cells from a research embryo initiate the circumstances ensuring that the loss would occur in any case? (Note that the 'loss' must be the fact that the embryo *dies*, as it is the fact that discarded IVF embryos will *die* in any case which, according to defenders of the discarded–created distinction, justifies extending the nothing-is-lost argument to the destruction of discarded IVF embryos). It cannot be that the researcher ensured that the embryo would die anyway by *creating* the embryo for the purpose of research, as the researcher who destroys the embryo does not necessarily have to be the one who created it. Yet, according to defenders of the discarded–created distinction, the researcher who 'merely' destroys an embryo created by another researcher cannot appeal to the nothing-is-lost argument either. Moreover, since it is not the creation of the research embryo which causes the certain death of the embryo, it cannot just be its creation which will cause the loss to occur anyway.

Perhaps a better candidate is the fact that the researcher who destroys the embryo *requested* the creation of the embryo. Through this request, the researcher initiated the creation of the embryo in circumstances in which the embryo was certainly going to be used for research and not for reproduction. This would be enough to rule out the nothing-is-lost argument as a justification for performing research on embryos whose creation one requested. However, not all those deriving stem cells from research embryos may have requested the creation of the particular embryo they are deriving stem cells from. So it must be something other than the request to create a particular embryo that ensures the embryo is bound to die from the start.

How, then, can we justify Outka's claim that any researcher deriving stem cells from a research embryo has initiated the circumstances ensuring that the loss would occur anyway? I think there is only one plausible justification. Such a researcher is typically part of an organised group of stem cell researchers whose shared intentions, agreements, and commitments make it the case that each embryo created for research will effectively be used for research, or at least will not have any prospect of survival. For there to be any prospect of survival, stem cell researchers would have to abstain from conducting research on the embryo, the embryo would need to have the biological potential to develop into a baby, and women would need to volunteer to carry the embryos to term. Given the commitments of the stem cell researchers, the fact that SCNT embryos do not possess the potential to become a baby at the moment, and the burden on women to carry embryos to term, this is extremely unlikely to happen. By being part of a particular group of stem cell researchers committed to research on stem cells from research embryos, the researchers conducting such research are part of the group collectively responsible for the creation of this embryo in circumstances that ruled out its survival. If the researcher suddenly abstained from the derivation of stem cells from this embryo, it would not be saved. It would either be used by another researcher, or would perish. It would die in any case. However, the researcher cannot appeal to this fact to justify killing the embryo. After all, she is part of a group that initiated the circumstances which ensure that this loss will occur. She shares responsibility for the embryo's inevitable death.[b]

To clarify why the nothing-is-lost argument is not credible in a situation where one is partly responsible for the loss to occur anyway, consider a hypothetical scenario in which Josef Mengele had attempted to justify his evil deeds by saying: "Well, Jews were going to be gassed anyway, so it was better to conduct fatal experiments on them that could potentially save other innocent lives; it didn't change their final disposition — premature death — so nothing was lost!" Indeed, part of the problem here is that Mengele, being a Nazi, was partly responsible for the fact that these prisoners were bound to die anyway.

To illustrate with another hypothetical example — one that is more analogous to the use of research embryos — suppose I work on an animal farm. My job is to breed and provide animals to my colleagues, who will then slaughter them for meat. It is plausible that to justify the slaughtering, my colleagues cannot appeal to the fact that the animals were going to die anyway. By being part of an organised group of people that provides and slaughters animals, the person who slaughters the animal is partly responsible for the fact that the animal existed in circumstances that made it virtually impossible for the animal to escape its destiny: being slaughtered. So where does this leave us?

---

[b] See Chapter 7 of this volume for more on arguments relating to this.

It seems that the nothing-is-lost argument cannot apply when one has initiated the circumstances which ensure that the loss will occur in any case, or is part of a group that is collectively responsible for having initiated these circumstances. By being part of a group enabling the creation of embryos whose only prospect is to die in research, one helps initiating the circumstances which ensure that the loss — the death of the embryos — will occur anyway. It has been argued that by deriving stem cells from discarded IVF embryos, one is not responsible in any way for the circumstances ensuring that the loss will occur in any case. Such embryos exist, and are condemned to death, because they are created to spare women from further egg harvesting and are unwanted for reproduction. Their existence is entirely separate from their use in stem cell research,[17] and it is therefore perfectly consistent, according to defenders of the discarded–created distinction, to accept the use and derivation of stem cells from discarded IVF embryos on the grounds that nothing is lost, while rejecting IVF practices that are responsible for the loss that would occur in any case.[18]

4.2.2 *Initiating Circumstances Ensuring that Future Losses will Occur in Any Case*

This cannot be the whole story, however. According to the discarded–created distinction, it is impermissible to *use* stem cells from research embryos even when those embryos were created by researchers with whom one has no close connection, for example, stem cell researchers abroad who routinely create research embryos and derive stem cells from these embryos for their own research. It would be difficult to argue that those who use stem cells that were derived from such embryos are partly responsible for the fact that these particular embryos were going to die anyway. So why does the nothing-is-lost argument not justify using stem cells in such cases?

At this point, the proponent of the discarded–created distinction could argue that, although using stem cells from research embryos that were not destroyed on request by the one who uses the stem cells need not have initiated the circumstances ensuring that the embryo would have died in any case, it does contribute to such circumstances for future research embryos. After all, by using these stem cells, one creates a demand for more research embryos and their stem cells, thereby initiating their existence in circumstances that close off any destination for the embryo but death. Even if the demand is not significant, and perhaps does not have a major impact on the number of research embryos created, by using stem cells from research embryos, each stem cell researcher contributes somewhat to further creation and destruction of research embryos.

So perhaps another reason why the nothing-is-lost argument may not apply is that although one does not at all contribute to the circumstances ensuring that the loss would occur in any case in the case in question, one contributes to such circumstances in the future. One is partly responsible for the fact that *something* is lost that would not have been lost otherwise. That something is not the life of the embryo from which the stem cells one is using were derived; it is the life of some future embryo.

Let us grant that this appeal to (circumstances ensuring) future embryo deaths can explain why something is lost in the case of the use of stem cells from research embryos. The question now arises whether a similar argument can be used to show that something is also lost when one performs research on stem cells from *discarded* IVF embryos. Adherents of the discarded–created distinction must deny this. They have attempted to do this by arguing that using stem cells from discarded IVF embryos does not create a demand for such embryos. Such embryos exist not for stem cell research but to protect women from the harms associated with egg harvesting. Thus, it is argued, discarded IVF embryos exist and will continue to exist on a large scale regardless of whether hESC research continues. The situation has been compared with that in the 'murder victim case':[19]

> *Murder victim case*: A teen has been murdered in gang violence. After having obtained the consent of the teen's parents, a surgeon uses the murder victim's organs for transplantation into a patient who requires them in order to survive.

The surgeon is not in any way responsible for the loss (the death of the teen). It is also extremely unlikely that the surgeon, by using the organ, will be responsible for similar future losses (more gang murders). Thus, not only is nothing lost by the murder victim, no loss is caused to anyone else either. Moreover, someone will be saved. It seems that the surgeon's actions are ethically acceptable for the reason that nothing is lost. If using discarded IVF embryos to obtain stem cells is similar to using murder victims for their organs, then we should judge that it, too, is acceptable for the reason that nothing is lost, even if we believe embryos have significant moral status or the same status as a typical adult human.

But is the Murder Victim Case analogous to the use and derivation of stem cells from discarded IVF embryos? If not, we cannot just infer from the comparison that nothing will be lost by using and deriving stem cells from discarded IVF embryos. In the following subsection, I argue that destructive research on IVF embryos is disanalogous to the Murder Victim Case, and in such a way that it cannot be justified by appeal to the nothing-is-lost argument. Thus, I will argue that, just as something is lost in the case of the use of stem cells from research embryos, so too it is lost in the case of the use and derivation of stem cells from discarded IVF embryos.

4.3 Something will be Lost

4.3.1 *First Disanalogy*

One apparent difference between performing destructive research on discarded IVF embryos and the Murder Victim Case is that, in the former case, it is the stem cell researcher who kills the embryo, whereas in the Murder Victim Case it is the gang, not the surgeon, who killed the teen. However, some have denied this difference and have argued that it is the parents who decide to donate their IVF embryo for research who kill it; the researcher only determines the manner in which it is killed, which is irrelevant as early embryos cannot experience pain.[20] It seems, however, hard to deny that the researcher at least shares responsibility in killing the embryo. That there are embryos left over after IVF treatment is one thing, the decision to use and destroy these embryos in research is another. Dan Brock has argued that this is why what he calls the 'strong version' of the nothing-is-lost argument, which appeals to the fact that the loss is going to occur anyway whatever anyone does, cannot apply to the destruction of discarded IVF embryos.[21] Brock points out that discarded IVF embryos are frozen, but still alive, and retain the biological potential to develop into a baby. That the embryos are going to die soon in any case is therefore not true. It is the decision to use these embryos for research that removes their biological potential, as well as any chance that it will be realised. Thus, there is a loss when one destroys a discarded IVF embryo.

One reply could be that, although it is true that the discarded IVF embryo is not dead but frozen, the difference is unimportant. If discarded embryos are not destroyed in research, they will instead be eternally stored in nitrogen freezers and nothing will be lost by denying embryos a life in these freezers.[22] This is what Brock refers to as the 'weaker version' of the nothing-is-lost argument. It looks at what will happen to the embryo given what others will *in fact* do, not what they *could* do. Indeed, these frozen embryos will not *in fact* be used for reproductive purposes, so, in this sense, nothing will be lost by using them for research. However, Brock argues that although the weaker version of the nothing-is-lost argument applies to the destruction of discarded IVF embryos, it should not be accepted by those who think the embryo has significant moral status. Accepting that nothing is lost because the embryo will not be 'rescued' by anyone in any case is like accepting the killing of an abandoned baby for its organs because it was going to die in any case. Brock argues that the right response in this situation is to do whatever one *could* do to save the baby, regardless of what other people will *in fact* do. He further concludes that "those who rightly reject this weaker version of the nothing is lost principle will argue that likewise, the alternative to destroying spare embryos in hESC research is to keep them alive and frozen or to give them to others for implantation."[23]

However, the defender of the discarded–created distinction could still object that although it is true that these embryos have the potential to live and could be rescued, it is not feasible to rescue all of them given the realities of IVF practices in many countries. In the US alone, about 400,000 embryos are stored in freezers. Nightlight Christian Adoption's 'Snowflakes program' has arranged adoptions of some of these embryos, but, despite their efforts to 'rescue' frozen embryos, the great majority are discarded. So adherents of the discarded–created distinction may argue that, although in an ideal world they would save all frozen embryos or prevent their existence, this is not possible now. They can then argue that if we cannot save the hundreds of thousands frozen embryos worldwide, we should not let them go to waste but should use them for beneficial research. Nothing will be lost.

So even though it is true that the embryos have the biological potential to become babies, if this potential cannot be realised even if one does whatever one can to realise it, then perhaps nothing important is lost by depriving them of that potential by destroying them in research. Thus, while it is true that stem cell researchers actually kill discarded embryos, whereas the transplant surgeon does not kill the teen, it seems doubtful that this difference is morally significant. The stem cell researchers do deprive the embryos of potential, but it is not clear that this is important given that the potential was extremely unlikely to be realised, despite efforts to do so.

4.3.2 *Second Disanalogy*

There is, however, another difference between the Murder Victim Case and the use of discarded IVF embryos in stem cell research — a difference that is more clearly significant. In discussing whether the use of tissue from aborted foetuses is analogous to the use of organs from murder victims, Lynn Gillam distinguishes three mechanisms by which using the murder victim's organs might result in more murders: (1) by changing society's moral beliefs about murder, (2) by decreasing the state's efforts to deter it, and (3) by strengthening incentives to commit murder.[24] She suggests that because condemnation of murder is so deep-rooted in our moral psyche and is strongly reinforced by law worldwide, it is very unlikely that any of these effects will occur if we use organs from murder victims; it is therefore ethically acceptable to do this. Yet the circumstances of abortion differ from those of criminal murder. Condemnation of abortion is not deeply-rooted in almost everyone's intuitions, abortion legislation differs from country to country and has changed over time, and abortion is generally performed by the medical profession within a regulatory framework. These circumstances make it more likely that lobbying from pro-choice groups and the potential benefits of foetal tissue research will provide an incentive to relax existing abortion regulation or will result in reduced efforts to prevent abortion. In such a climate, it may also be easier for women to justify their decision to abort their foetus. Gillam's point is that the differences in circumstances between abortion and criminal murder make it more likely that foetal tissue research will result in an increase in the number of abortions performed than that transplantation from murder victims will increase the murder rate.

Gillam's reasoning also applies to hESC research as the circumstances are very similar to those of foetal tissue research.[25] Using discarded IVF embryos can, through mechanisms similar to those described by Gillam, indirectly promote embryo destruction. For example, the benefits of hESC research may weaken efforts to reduce the number of embryos discarded in IVF. Although a society that accepts IVF will not suddenly stop offering IVF treatments, it might still change the practice in such a way that fewer surplus embryos are created, and thus fewer embryos will die. By routinely using discarded IVF embryos in research, one removes the incentive to develop forms of IVF that result in fewer discarded embryos. Moreover, because hESC research has the potential to benefit many more people than foetal tissue research and the moral status of early embryos is more controversial than that of older foetuses, restrictions on the use of embryos for research and therapies (including fertility treatment) might be more easily loosened than the abortion regulations that Gillam has in mind. Using discarded IVF embryos may also cause those already uncertain about the moral status of the embryo to feel increasingly comfortable about the use of embryos for beneficial research, and even about creating them for research.

Perhaps defenders of the discarded–created distinction are right that the use of discarded IVF embryos to obtain stem cells does not create a demand for discarded IVF embryos. But there is another way in which it might, indirectly, promote embryo destruction due to IVF: By altering our attitudes to it. If embryos have significant moral status and it is true that one indirectly promotes embryo destruction by using discarded IVF embryos, it seems that something significant will be lost; though it may be true that nothing-is-lost *for the very embryos that the researchers in question destroy*, doing the research may contribute to losses suffered by other embryos in the future.

According to the discarded–created distinction, a researcher using or deriving stem cells from a research embryo cannot appeal to the nothing-is-lost argument because he is partly responsible for the fact that the embryo was going to die in any case, or for the fact that some future embryo is going to die anyway. However, I have now argued that a similar result holds in the case of using and deriving stem cells from discarded IVF embryos. Even though it may be true that this research does not contribute to the unavoidable death of the embryos one destroys, it does contribute to the unavoidable death of future embryos, albeit via different mechanisms. In neither case, then, is the nothing-is-lost argument persuasive. Adherents of the discarded–created distinction must convincingly show that, although deriving stem cells from research embryos contributes to future embryo deaths, deriving stem cells from discarded IVF embryos does not. Unless they can do this, the nothing-is-lost argument cannot support the discarded–created distinction.

5. A FINAL ATTEMPT TO SAVE THE DISCARDED–CREATED DISTINCTION

5.1 Using Research Embryos Contributes to a Greater Loss

Perhaps there is another argument that can support the discarded–created distinction. Suppose we accept that both research on discarded IVF embryos and research on research embryos contribute to future losses; the former contributes to future IVF practices that result in many embryo deaths, and the latter to increased demand for, and thus further deaths of, research embryos. Still, it might be argued that the latter losses are more important than the former. Moreover, the difference in costs (not the direct cost of deriving the stem cells but downstream costs of further embryo deaths) could be sufficient to ensure that the benefits of stem cell research outweigh the downstream costs of doing research on discarded IVF embryos, but do not outweigh the downstream costs of doing research on research embryos. This could be so, for example, if it is significantly worse that embryos die for research than for reproduction (as in IVF).

5.2 Worse that Embryos Die for Research than for Reproduction

Some have argued it must be significantly worse that embryos die for research than for reproduction, as in IVF. Two reasons could be advanced to support this claim: (1) research embryos are treated as a mere means whereas discarded IVF embryos are not (or not to the same degree), and (2) discarded IVF embryos die in the service of more important goods than do research embryos. Let us take a closer look at each of these arguments.

5.2.1 *Research Embryos are Treated as Mere Means*

It has been argued that by creating research embryos, we treat the embryo as a mere means, thereby not handling it with the appropriate respect such a form of human life is entitled to. The underlying idea is that respect for human beings prevents the instrumental use of embryos, an act which according to some violates human dignity.[26] However, embryos are used as a mere means in IVF too.[27] As mentioned earlier, IVF — as it is practiced in most countries — involves the deliberate creation of many surplus embryos that will be discarded. The reason for creating such a surplus of embryos is not the hope that each of the embryos will develop into a baby. It is to spare women from further egg retrievals. Embryos created in IVF are thus mere means too.

In response to this observation, it is often objected that IVF embryos are created with the *primary* intention to create a child. The principle of double effect is then invoked, and it is argued that the fact that remaining IVF embryos are discarded on a large scale is merely an unintended side-effect of IVF and therefore acceptable. However, it seems implausible to speak of a 'side-effect' when we know that for every embryo transferred to the womb, another will be discarded (In the UK, for example, more than two million IVF embryos were created between 1991 and 2005, more than half of which were discarded).[28] In any case, even if creating surplus embryos is a side-effect, it is not an unintended one. After all, IVF could be done without creating surplus embryos. In some countries, IVF does not involve the creation of surplus embryos. Creating surplus embryos is not an unavoidable side effect of IVF. It is intentionally brought about in order to reduce harm to women undergoing IVF.

Yet if the instrumentalisation of embryos to help reduce harm to women undergoing IVF is permissible, why then is the instrumentalisation of embryos for important biomedical research and for developing treatments for currently incurable diseases impermissible? One possible reply is that IVF involves a different kind of instrumentalisation. At the time of creation, IVF embryos have a prospect of implantation, even if once not selected for implantation, they will be discarded. Research embryos do not have this prospect. They are created for a purpose entirely external to themselves and are therefore used as mere means to a greater degree, or in a more problematic way.[29] Is it really the case that having the prospect of further development makes a significant moral difference? This seems implausible. Consider the following scenario. Suppose we create research embryos, but randomly select some of these embryos for donation to infertile couples. At the time of their creation, research embryos, just like IVF embryos, would have the prospect of developing into a baby. This would be true even if we donated only one embryo to an infertile couple. This does not seem to treat the embryos with more respect. A better candidate for treating them with respect seems to be to restrict the use of embryos to important scientific research (like we do with cadavers or animal research). Note that only a minority of left-over IVF embryos is donated for research; most left over embryos are just left to perish. If respect for the embryo is what is so important, it is not clear how contributing to letting IVF embryos perish on a large scale is more respectful than destroying embryos for, and only for, research that can save and improve many people's lives.

5.2.2 *Discarded IVF Embryos Die in the Service of More Important Goods*

Some people may argue that the goods produced by IVF are more important than the goods produced by using research embryos. Permissive attitudes to (and regulations regarding) the destructive use of embryos in the context of IVF have mostly been justified by reference to the right to reproductive liberty, which in turn has been justified by pointing to the central position of reproductive projects in people's lives. These justifications cannot be invoked to defend the use of embryos in the context of research.[30] Although this is true, it seems implausible that protecting reproductive liberty and helping infertile people to realise their wish for a child produces more important goods than saving people's lives. After all, it is generally accepted that saving lives is more important than creating lives. It is also plausible that saving lives is more important than protecting reproductive liberty.

There are other reasons why the goods produced by IVF could be more important. One may argue that by not creating surplus embryos, we harm women undergoing IVF, whereas by not creating research embryos, we merely allow harm to occur to individuals in need of stem cell based therapies. Many believe that it is typically worse to do harm than to allow harm to occur. For example, many believe that it is typically wrong to kill a person even in circumstances in which it would be permissible to let that person die. However, although it may be true that not creating surplus embryos causes at least some of the harm to women undergoing IVF (as they may have to undergo hormone treatment several times), it is mainly the women's decision to proceed with IVF which harms them. Not creating surplus embryos can then be seen as 'merely' allowing harm to occur and consequently is, morally speaking, on a par with not creating research embryos. The 'doing versus allowing harm' distinction, insofar as it has any force in this context, cannot provide a ground for arguing that the goods produced by using discarded IVF embryos are greater than those produced by using research embryos.

To support the claim that creating surplus embryos in IVF produces more important goods, one may also argue (1) that we should care more about current harm to present individuals than about future harm to them, or (2) that we should give more weight to harm to current individuals than to future individuals. However, first of all, it seems implausible that mere location in time accords special significance to a benefit or harm. If my neighbour burns painted and glued wood in his chimney on a daily basis, and as a result of the poisonous smoke I get lung cancer, it should not matter morally whether I get cancer now, or within 10 years. In both cases, my neighbour should be stopped from burning treated wood now. Secondly, there does not seem to be a strong reason for an institution to treat the interests of future individuals any differently from those of present individuals either. Suppose the poisonous particles remain in the air for more than 20 years, and an individual that was not born at the time the burning took place now lives in my apartment and gets lung cancer as a result of the burning of treated wood 20 years ago. We have as much reason to stop my neighbour from burning wood now. After all, it is equally bad for the future person to get lung cancer than for a currently living person. The 'badness' consists in getting the cancer, not in what generation it occurs (assuming the circumstances are similar).[31] Likewise, the fact that the women undergoing IVF exist now does not make the goods produced by killing embryos for IVF more important than the goods produced by killing embryos to prevent harm to future persons in need of stem cell-based treatments.

6. CONCLUSION

I started this chapter by investigating how proponents of the discarded–created distinction justify the destruction of discarded IVF embryos to obtain stem cells and showed that this justification also applies, at least at first sight, to the destruction of research embryos. This suggests that there should be a presumption against the discarded–created distinction. I then considered whether this presumption can be overridden. I concluded that the strongest arguments adduced in support of the discarded–created distinction do not hold. The nothing-is-lost argument fails. Whether one performs stem cell research using discarded IVF embryos or research embryos, one contributes to future embryo deaths and, thus, something is lost. The argument that the moral (downstream) costs of the two types of stem cell research are sufficiently different that, in the one case they are outweighed by the benefits, but in the other they are not, does not hold either. In both IVF and embryo research, embryos are clearly treated as a mere means. Moreover, it is implausible that the goods created through IVF are more important than the goods created by experiments with research embryos. Saving lives is more important than creating lives, and both IVF and stem cell research aim to prevent harm. It does not matter whether this is current or future harm, or harm to existing or future people. It seems, then, that once one accepts the use and derivation of stem cells from discarded IVF embryos, it is hard to justify rejecting the use and derivation of stem cells from research embryos. Since there is, I have argued, a presumption against the discarded–created distinction, as long as no good argument has been adduced in support of the discarded–created distinction, we should reject it.

REFERENCES

1. National Advisory Bioethics, Commission. (1999) *Ethical Issues in Human Stem Cell Research.* NBAC, Rockville, MD.
2. National Advisory Bioethics, Commission. (1999) *Ethical Issues in Human Stem Cell Research.* NBAC, Rockville, MD.
3. Commission of the European Communities. (2003) *Report on Human Embryonic Stem Cell Research.* European Commission, Brussels, p. 9.
4. National Advisory Bioethics, Commission. (1999) *Ethical Issues in Human Stem Cell Research.* NBAC, Rockville, MD, p. 52.
5. National Advisory Bioethics, Commission. (1999) *Ethical Issues in Human Stem Cell Research.* NBAC, Rockville, MD.

6. Fletcher JC. (2001) NBAC's Arguments on Embryo Research: Strengths and Weaknesses. In: Holland S, Lebacqz K, Zoloth L (eds), *The Human Embryonic Stem Cell Debate: Science, Ethics, and Public Policy.* MIT Press, Cambridge, MA, pp. 61–72.
7. Pennings G, Van Steirteghem A. (2004) The Subsidiarity Principle in the Context of Embryonic Stem Cell Research. *Human Reproduction* **19(5):** 1060–1064.
8. McHugh PR. (2004) Zygote and 'Clonote': The Ethical Use of Embryonic Stem Cells. *New England Journal of Medicine* **351:** 209–211.
9. Outka G. (2009). The Ethics of Embryonic Stem Cell Research and the Principle of Nothing is Lost. *Yale Journal of Health Policy & Ethics* **9:** 585–602.
10. Ramsey P. (1961) *War and the Christian Conscience: How Shall Modern War be Conducted Justly?* Duke University Press, Durham, NC.
11. Prieur MR, Atkinson J, Hardingham L, *et al.* (2006) Stem Cell Research in a Catholic Institution: Yes or No? *Kennedy Institute of Ethics Journal* **16(1):** 73–98.
12. Pennings G, Van Steirteghem A. (2004) The Subsidiarity Principle in the Context of Embryonic Stem Cell Research. *Human Reproduction* **19(5):** 1060–1064.
13. Prieur MR, Atkinson J, Hardingham L, *et al.* (2006) Stem Cell Research in a Catholic Institution: Yes or No? *Kennedy Institute of Ethics Journal* **16(1):** 73–98.
14. Zoloth L. (2002) Jordan's Banks: A View from the First Years of Human Embryonic Stem Cell Research. *American Journal Bioethics* **2:** 3–11.
15. Ramsey P. (1961) *War and the Christian Conscience: How Shall Modern War be Conducted Justly?* Duke University Press, Durham, NC, p. 598.
16. Ramsey P. (1961) *War and the Christian Conscience: How Shall Modern War be Conducted Justly?* Duke University Press, Durham, NC, p. 596.
17. Robertson JA. (2004) Causative vs. Beneficial Complicity in the Embryonic Stem Cell Debate. *Connecticut Law Review* **36:** 1099–1113.
18. Ramsey P. (1961) *War and the Christian Conscience: How Shall Modern War be Conducted Justly?* Duke University Press, Durham, NC.
19. Green RM. (2002) Benefiting from 'Evil': An Incipient Moral Problem in Human Stem Cell Research. *Bioethics* **16:** 544–56.
20. Green RM. (2002) Benefiting from 'Evil': An Incipient Moral Problem in Human Stem Cell Research. *Bioethics* **16:** 544–56.
21. Brock DW. (2010) Creating Embryos for Use in Stem Cell Research. *Journal of Law & Medical Ethics* **38(2):** 229–37.

22. Outka G. (2009). The Ethics of Embryonic Stem Cell Research and the Principle of Nothing is Lost. *Yale Journal of Health Policy & Ethics* **9:** 585–602.
23. Brock DW. (2010) Creating Embryos for Use in Stem Cell Research. *Journal of Law & Medical Ethics* **38(2):** 229–37, p. 232.
24. Gillam L. (1997) Arguing by Analogy in the Fetal Tissue Debate. *Bioethics* **11(5):** 397–412.
25. Takala T, Häyry M. (2007) Benefiting from Past Wrongdoing, Human Embryonic Stem Cell Lines, and the Fragility of the German Legal Position. *Bioethics* **21(3):** 150–159.
26. Heinemann T, Honnefelder L. (2002) Principles of Ethical Decision-making Regarding Embryonic Stem Cell Research in Germany. *Bioethics* **16:** 530–43.
27. Savulescu J. (2004) Embryo Research: Are There Any Lessons from Natural Reproduction? *Cambridge Quarterly of Healthcare Ethics* **13:** 68–95.
28. Woolf M. (December 30, 2007) IVF clinics destroy 1m 'waste' embryos. Available at http://www.timesonline.co.uk/tol/life_and_style/health/article 3108160.ece (Accessed 16 June 2011).
29. House of Lords Select Committee on Stem Cell Research. (2002) *Stem Cell Research*, Section 4.27.
30. Holm S. (2003) The Ethical Case Against Stem Cell Research. *Cambridge Quarterly of Healthcare Ethics* **12(4):** 372–83.
31. Savulescu J. (2007). Future People, Involuntary Medical Treatment in Pregnancy and the Duty of Easy Rescue. *Utilitas* **19(1):** 1–20.

7

Stem Cell Therapies & Benefiting from the Fruits of Banned Research

Muireann Quigley

Despite the medical promise that stem cell research offers, human embryonic stem cell (hESC) research remains morally problematic for some. The putative moral wrong inherent in such research is the destruction of embryos. This moral position is reflected in the legislative responses of some countries which have banned hESC research. This chapter considers whether it would be morally wrong for those countries, which ban human embryonic stem cell research on the grounds of embryo destruction, to benefit from hESC-derived therapies once they are available. Determinations of complicity in stem cell research are difficult to make because we run into problems of collective action and responsibility. Nonetheless, here it is argued that the morally conscientious and consistent State would not import hESC-derived therapies for use within their borders where that State persists in maintaining that embryo destruction is morally impermissible. Recognising that there might be a concern that such countries are free-riding on the stem cell research efforts of other countries, this chapter also briefly examines whether it would be justifiable for third parties organisations or countries to take steps to prevent them from importing hESC-derived therapies.

Centre for Social Ethics & Policy, Institute for Science, Ethics, & Innovation, School of Law, University of Manchester, Oxford Rd., Manchester, M13 9PL, United Kingdom.
Email: muireann.quigley@manchester.ac.uk

> The wrongs are going to be committed anyway; there is nothing to be done about that. But one must not be implicated in them. One's own hands must remain clean even as the world falls.[1]

1. INTRODUCTION

Imagine if you will, that some time in the not too distant future, researchers have developed a very effective treatment for type 1 diabetes.[a] Due to this treatment, diabetes patients no longer have to inject insulin several times a day, nor do they suffer from the many long-term effects (heart disease, stroke, renal failure, blindness) that diabetes can cause. This new treatment does not merely ameliorate the disease and its effects, or simply alleviate its symptoms; it actually cures the patient of the disease. It does this by completely restoring the function of a patient's insulin-producing pancreatic islet cells.[b,2] Those with type 1 diabetes have a mortality rate 5.6 times higher than the general population.[3] As such, it is estimated that this new treatment will save thousands of lives each year.[c,4] In addition, it is expected that the morbidity associated with type 1 diabetes will decrease dramatically for those who elect to have this treatment.

However, for those of you who are ethically-minded, there might be something of a problem with this new treatment. It is based on previous

---

[a] Type 1 diabetes mellitus (insulin-dependent diabetes mellitus) is characterised by the destruction of the pancreas' insulin-producing cells, leading to a deficiency in insulin. This in turn causes a rise in blood glucose levels as the glucose cannot be transported into the body's cells in the absence of insulin. In type 2 diabetes mellitus (non-insulin-dependent diabetes mellitus) insulin is still produced, but the body's cells are resistant to it, also leading to a rise in blood glucose levels.

[b] For example, a clinical trial in progress at Uppsala University Hospital in Sweden is evaluating the outcome of treatment type 1 diabetes with mesenchymal stem cells — see note 2.

[c] In 2007 in the United States alone, diabetes was recorded as the cause of death on 71,382 death certificates. Although this includes those with type 1 and type 2 diabetes; with type 1 diabetes accounting for an estimated 5% of cases this would account for 3,569 deaths in that period — see note 4.

research with human embryonic stem cells (hESCs). Without such research, the treatment could not have been developed and would not exist. Human embryonic stem cell research necessarily involves the destruction of embryos and, in the course of the research leading up to and in developing this new treatment, many embryos have been destroyed. Likewise, until new cell lines from which the cells for treatment are derived are no longer needed, the production of the treatment itself means that embryos are destroyed. The research team which has developed this treatment, the pharmaceutical company who is manufacturing it, and the government of the country where it was developed are unusually morally conscientious. They recognise the value of pluralism that exists both within and between countries; that different countries and their citizens occupy differing moral positions with regards to research involving human embryos. In particular, they note that some countries have been morally opposed to hESC research in the past and have, for moral reasons, chosen to ban hESC research. For this reason, they are concerned that these countries, and indeed their citizens, might inadvertently become complicit in what they themselves deem to be the moral wrongdoing of hESC research.

Despite the medical promise of this development and other research involving hESCs, it is the fact of embryo destruction that is likely to be morally problematic for many. This is reflected in the regulatory responses of some countries within Europe, and elsewhere, to hESC research.[5] The moral crux of the matter for these countries (and perhaps their individual citizens) is the moral status of the embryo. For them, the human embryo has a special status worthy of protection and, for that reason, any research involving the destruction of the human embryo is considered to be morally wrong. However, if this is correct and hESC research is morally wrong, would it be morally wrong to benefit from or use hESC-derived therapies once they are available? Would doing so render those countries which import such therapies complicit in the immoral acts that brought about the development of such therapies in the first place? This paper offers an examination of these questions and the arguments regarding stem cell treatments and complicity in hESC research.

2. STEM CELLS, COMPLICITY, AND MORAL WRONG-DOING

The starting premise of this paper is that it is morally wrong to be complicit in immoral acts. If an act in itself can be said to be morally wrong, then it is also morally wrong to 'aid and abet' others in the performance of this act. By becoming an accomplice in an act, an individual, organisation, or state elects to share in the moral responsibility for that act and its consequences. Where the act is morally wrong the particular individual, organisation, or state must take their share of the moral blame. Therefore, where the putative moral wrong is the destruction of human embryos for stem cell research, a country which bans hESC research based upon its moral opposition to said destruction ought not to be complicit in the wrongs contained therein. This is the case even if those moral wrongs are brought about by other people or countries. As such, this means not only refraining from providing the stem cell scientists with the resources and tools to carry out their research, but might also entail refraining from profiting or benefiting from those immoral actions once they have occurred.

For the sake of argument, let us assume that there is a country out there which, having banned hESC research, actually has coherent legislation regarding such research. This country, let's call it Fantasia, does not allow the creation of human embryonic stem cell lines within its borders. As such, it does not allow the creation of embryos purely for research, nor does it allow the use of spare IVF embryos for that purpose. It does not even permit the import of hESC lines from other countries which sanction such research.[6] Additionally, while it provides a limited infertility service to its citizens, it allows only the creation of embryos which will be implanted. Thus, Fantasia appears to have taken all reasonable measures to protect against the destruction of embryos and seems able to claim consistency and 'clean hands' from this point of view. In spite of this, are there other ways in which it might become complicit in the moral wrongs that it purports lie in hESC research? Would the import and use of hESC-derived therapies such as the one in the scenario at the start of this paper make Fantasia causally complicit in hESC research? In order to answer this, we need to first look at the ways in which these countries might be considered to be complicit in the moral wrong.

Gardner maintains that "accomplices themselves bring wrongdoing into the world", and that "an accomplice is one who acts with the consequence or result that the principal commits the wrong".[7] There could be two main ways in which an individual could be complicit in the wrongdoing of another. The first would be to intentionally influence the actions of the wrongdoer, and the second would be to help the wrongdoer to commit the wrong, but where he does not actually participate in the wrongful activity.[d,8] The second of these is easier to deal with. In order to do so, let us consider the following example.

If an individual, Peter, supplies a gun to his colleague, Jane, knowing that she is going to use it to shoot her boss, Fred, then Peter must share in the moral culpability for that death. It is clear here that he is morally responsible in some way for Fred's death. This is the case even if we decide that he does not share the same degree of responsibility as Jane herself. Here, Peter aids and abets Jane in committing the moral wrong, even though he does not participate in the act himself. Of course, it is not always obvious exactly what participation in morally dubious acts consists of. Is Peter culpable if he was not aware of the use Jane would put the gun to? What if he claimed he knew of the gun's intended use but he did not intend Fred's death? And what if he both knew about the gun's intended use and intended Fred's death, yet Jane is a terrible shot, therefore, not only missing Fred, but having little real chance of actually hitting him? The purpose of asking these questions is not to suggest that culpability is not present in these situations, but simply to highlight that the notion of moral culpability is complex; it ought not to be seen as an all or nothing concept and may be present in degrees.

An initial way of looking at this in the context of hESC research is to view the stem cell researcher who obtains the inner cell mass from the embryos as the primary source of any wrongdoing. This is because it is his direct actions that bring about the destruction of those embryos. In

[d] These are taken directly from Stanley Kadish's thesis regarding criminal liability — see note 8.

addition, others will be causally implicated in the destruction of the embryos if they knowingly help him. Therefore, if you supply the pipette used by the stem cell scientist to remove the inner cell mass of the embryo, knowing what it will be used for, you are in some way morally responsible for the ensuing destruction of that embryo. Here, one might be tempted to invoke something akin to the arms dealer defence and claim that if you had not supplied the pipette, someone else or some other company would have. Gardner suggests that there will be times when, although a particular contribution is not necessary for the wrong to take place that, from the (potential) accomplice's perspective, it does make a difference to the overall incidence of wrongdoing in the relevant sense.[9] For him, participation in or encouragement of wrongdoing is not a zero-sum game. One cannot offer as a defence the fact that the wrong would have been committed regardless, or that others would be complicit in it even if a particular individual is not. To do so is to abdicate responsibility for one's own (in)actions. In the case of the arms dealer, Gardner notes:

> That someone else will do it means that the individual arms dealer makes *no difference* to whether the arms will be supplied. But it does not follow that the dealer makes no *causal* contribution to their being supplied.[10]

The reason is that by them supplying the arms rather than someone else, they have stepped into the causal chain. Likewise, although it is true that your supplying the pipette to the stem cell scientist makes no difference to whether or not that particular research will take place, it is not the case that there is no causal connection where you do. Thus, any contributory responsibility cannot be abdicated merely because someone else would have acted in your absence.

In simple examples such as the ones above, the causal link between the accomplice and the immoral act seems relatively uncomplicated and serves to demonstrate that "not all causal antecedents are necessary conditions of their consequents".[11] However, *benefiting* from treatments based on hESC research is not an example of this type of direct complicity. Nonetheless, it might be an example of indirect complicity, where the actions of the wrongdoer are *influenced* by the actions of those who import the hESC-derived therapies.[e,12] In order to examine this, I want to look at the arguments of two commentators who think that those who oppose hESC research can nonetheless benefit from the fruits of such research with moral impunity.

3. ARE hESC THERAPIES MORALLY TAINTED?

Two authors who argue this are John Robertson[13] and Ronald Green.[14] They both maintain that if individuals believe that hESC research is morally wrong, they can still benefit from it without fear of moral taint. They present similar, albeit slightly different, reasoning for this assertion. Let us look first at Robertson.

3.1 On Causing and Benefiting

Robertson contends that there is a "moral distinction between causing and benefiting from a moral evil".[15] Contrary to some opponents of hESC research, he argues that "the distinction is real and has moral weight".[16] If such a distinction holds, then it brings with it an advantage for those who might be troubled by using therapies derived from embryonic stem cells or consequent on research involving such cells. A robust distinction would mean that individuals could use those therapies without fear of the moral taint of being complicit in embryo destruction. The basis of the distinction, for Robertson, is that not all instances of benefiting from past actions will themselves have *caused* those actions to occur. Here he uses the example of receiving an inheritance after another's death. The mere

[e] Note that I am not going to examine the intentionality aspect of the claim that one is complicit where one intentionally influences the actions of the wrongdoer. However, this aspect is important for some commentators — see note 12 below.

fact that a person benefits from the death through receiving the inheritance does not make them causally complicit in that death.[17] He further illustrates his point with the so-called murder victim scenario, already mentioned by Katrien Devolder in the previous chapter of this volume. Here, Robertson maintains that the "distinction makes transplant of organs from murder victims morally acceptable even though murder is immoral and criminal".[18] As Devolder noted in relation to this:

> The surgeon is not in any way responsible for the loss (the death of the teen). It is also extremely unlikely that the surgeon, by using the organ, will be responsible for similar future losses (more gang murders). Thus, not only is nothing lost by the murder victim, no loss is caused to anyone else either. Moreover, someone will be saved. It seems that the surgeon's actions are ethically acceptable for the reason that nothing is lost.[19]

Thus in this scenario, we can see that even though the recipient benefits from the transplant of the organ, neither he nor the surgeon can be viewed as having caused the death or as being complicit in the death. We should, at this point, note that the analogy being drawn with the murder victim case is in relation to spare IVF embryos rather than embryos which have specifically been created for use in research. These are embryos which would otherwise be fated for destruction if not used in research.

Yet, as Devolder argued in her chapter, we can only judge the use of leftover IVF embryos as acceptable in the same manner that the transplant of organs from murder victims would be if the cases are actually analogous. There she argued that the cases are disanalogous in two important ways. First, unlike the surgeon in the murder victim case, the stem cell researcher at least shares in the responsibility for embryo death/destruction. Second, following Lynn Gillam's arguments regarding the use of foetal tissue in research, Devolder argued that "using discarded IVF embryos may also cause those already uncertain about the moral status of the embryo to feel increasingly comfortable about the use of embryos for beneficial research".[20] In this manner, because there is uncertainty regarding the morality of embryo use, where such certainty does not exist regarding the murder of fully fledged agents, the use of spare embryos for research could lead to an alteration in attitudes towards embryo destruction. Yet it is unlikely that the use of organs from a murder victim would change attitudes *vis-à-vis* the wrongness of murder in the same way.

The first of these points is important with respect to the issue of causation, and I will return to this later. For now, however, I want us to allow the possibility that not all instances of *benefiting* from hESC research can actually be said to have *caused* that research to take place in the first place. For example, one could argue that in a country which allows this type of research, prior to the development of any derivative therapies, the potential for benefits and therapies are part of the driving force behind the research in the first instance. Here one could argue that the possibility of the future benefit creates a demand which causatively contributes to the occurrence of the research. But countries such as Fantasia which ban such research are not involved in this kind of incentivisation; therefore, they cannot be said to be causally complicit. What is important here is the causative effect that our actions will have on future events. It is not logical to claim that the future benefit *causes* the prior event. As Heidi Mertes and Guido Pennings note, "complicity does not work retroactively".[21]

3.2 Moral Encouragement of Wrongdoing

However, while we might agree as a point of logic that complicity does not work retroactively, there will be scenarios where the (potential) future benefit does causally contribute to the putative moral wrong of embryo destruction. Once hESC-derived therapies have been developed and are available, there might be instances where the actions of countries such as Fantasia could be said to influence the moral wrongdoing if they choose to import those therapies. The wrong in this case would not be the wrong of being complicit in the research that has *already taken place*, but that the act of importing and using hESC therapies might lead to a continuance of such research and its associated practices. This has the effect that the wrongdoing occurs again and again in the future. Despite this, Robertson maintains that, in the case of hESCs, profiting or benefiting from the past wrongs does *not* 'encourage' future wrongs.[22] He says that this is because "what is perceived as a 'wrong' is legally permitted and will occur on a widespread basis whatever the decision of a particular jurisdiction".[23] Robertson's support for the causation/benefit distinction may in part be because he does not want to align himself with those who would oppose hESC research. Be that as it may, it is difficult to see how being complicit in the purported moral wrong can be construed as being morally right in itself *merely* because the principal wrong takes place in another jurisdiction. Nonetheless, let us take a generous interpretation of Robertson's argument and see it as the claim that, regardless of the actions of individual countries or jurisdictions, there will still be the same *amount* of the putative moral wrong in the world. That this is correct is open to debate, but we will proceed on the assumption that it is. At this point I want to turn to Green's arguments, since he proceeds on a similar basis.

Like Robertson, Green also asserts that individuals who oppose hESC research on the grounds of embryo destruction can benefit from it without incurring moral blame. Again, like Robertson, he is concerned with the 'moral encouragement' of the alleged wrongs.[24] He identifies three types of moral encouragement that would make an individual complicit in a moral wrong. These are (1) direct encouragement through agency — where an individual gets another to carry out the wrong on their behalf; (2) direct encouragement through acceptance of benefit — where an individual in some way knowingly profits from and overlooks the wrongdoing, thus encouraging it again in the future; and (3) indirect encouragement through legitimisation of a practice — by benefiting from the wrong a 'rule of conduct' comes into being which governs people's behaviour in the future.[25]

Green admits that hESC research appears to contravene at least two of these: encouragement through the acceptance of benefit and encouragement through legitimisation.[26] He claims, however, that "an understanding of the realities of hESC production" leads us to the "key insight" that "embryo destruction is entirely independent of hESC research and therapy".[27] The basis for this assertion is the fact that thousands of leftover IVF embryos are destroyed every year and, according to Green, this destruction occurs *independently* of any hESC research. He thinks that the causal link is missing, claiming that hESC research determines the *mode of destruction* of the embryos, not the destruction itself, as they were fated for this end regardless.[28] This does seem initially plausible and, on this argument, Fantasia's stance regarding hESC research will not actually make a difference to the number of embryos in the world that are destroyed (the alleged moral wrong). Therefore, they would not be morally culpable if they chose to import and use hESC-derived therapies. As we can see, this is analogous to Robertson's argument.

It seems that Green is appealing to the 'but-for' test of causation here.[29] The question he is essentially asking is, *but for* the hESC research, would the moral wrong, the destruction of the embryos, still have taken place? The answer, of course, is yes; the destruction of those same spare IVF embryos would still have occurred. Nevertheless, the 'but-for' test is not, and cannot be, the whole story when it comes to causation and moral responsibility. Consider if you will the principal actors in the following examples. In the first, Simon works at an IVF clinic, and it is part of his job to destroy the clinic's spare embryos. He does this by incineration of the embryos. In the second, Joan is a stem cell scientist, and as part of her job she extracts the inner cell mass from embryos; it is from this that a stem cell line is created. She extracts the inner cell mass after chemically dissolving the outer layer of cells; in effect destroying the embryo. Every so often, one of the IVF clinic's couples decides that, instead of having their spare embryos disposed of by the clinic, they will donate them for hESC research. In those cases, those embryos that would have been destroyed by Simon are actually destroyed by Joan. Green asserts that it is the *decision to discard* the embryos, not the hESC research, that causes their destruction.[30] If this is correct, the moral wrong of this lies with those who made that decision and those who were complicit in making the decision. For the sake of argument, this would minimally be the embryo's 'parents' and the IVF clinic. According to the reasoning implicit in Green's key insight, no moral blame will be apportioned to Joan (who actually carried out the act that destroyed the embryo), to the laboratory which carries out the hESC research, or to those who would benefit from such research, because those embryos *would have perished anyway*.[f]

---

[f] Of course, Green is only talking about spare embryos, not specially created ones.

This line of reasoning is intuitively appealing, yet there is something not quite right about it. Consider for a moment the following analogy: Peter has decided to shoot Jane, but, just as he is about to pull the trigger, Fred shoots Jane instead. Peter is a hit man who is working as an agent for the local mob leader. Whether or not he kills Jane is matterless, as the decision that Jane will die has been made. If Peter does not kill her, somebody else will. The mere fact that Jane would have died anyway because Peter would have shot her cannot absolve Fred of moral blame for that death; after all, he was the one who actually pulled the trigger. This is an example of a *novus actus interveniens* (a new intervening act).[31] This new act breaks the chain of causation rendering the original would-be wrongdoer not liable for the wrongdoing.[g] In the case of hESC research, the research itself is as a *novus actus interveniens*, rendering those who actually carry out the destruction of the embryos morally blameworthy. The hESC research is the proximate cause of the embryo destruction in this case.[h]

If hESC research can be considered to be the proximate, and thus the proper cause of the embryo destruction, this has implications for those who would benefit from this research. The action itself, the destruction of the embryo, is not performed by the persons who benefit from the treatment. The moral wrong itself is not committed by them since they do not engage in the wrong action. However, to quote Gardner, they do act "with the consequence or result that the principal commits the wrong".[32] In order for one to maintain that this is not the case, one has to believe, as Katrien Devolder and John Harris observe, that "the act through which the cell products are obtained should be completely separated from the use that is made of the products".[33] As such, where demand for hESC lines is being created through demand for hESC-derived therapies, claims of such separation would be difficult to maintain. This is because of the moral encouragement given by the

[g] Although this does not exclude the possibility that the original would-be wrongdoer might be complicit in the wrongdoing of the new wrongdoer; he simply cannot be considered to be the principal actor.

[h] Where embryos are created specifically for research there can be no question over the cause of the destruction.

beneficiaries of hESC-derived therapies to on-going hESC research. In this manner, contra Green's key insight, where countries such as Fantasia choose to import such therapies, they could be viewed as giving direct encouragement to the continued commission of embryo destruction by accepting the benefits of derivative therapies. Further, depending on the level of demand for these therapies, contra my interpretation above of Robertson's claim (that regardless of the actions of individual countries there will still be the same amount of the moral wrong in the world), the import of stem cell therapies by those countries which ban hESC research might in fact lead to increased embryo destruction.

3.3 Complicity through Benefit and the Problem of Collective Action

Thus far, the arguments seem to be leading to the conclusion that countries which oppose hESC research on the grounds of embryo destruction may not be able to avoid the taint of moral complicity if they import hESC derived therapies once available. Yet, the examination of complicity through benefit needs to go a little deeper. The reason is that in thinking about the notion of complicity in the context presented here, we run into two inter-related problems. The first is in determining a particular country's causal contribution to embryo destruction through the (direct or indirect) encouragement of hESC research (or indeed whether there is any causal contribution at all). The second is how this relates to the notion of *collective* action and responsibility. The difficulty, as Christopher Kutz notes, is that certain putative "harms and wrongs are essentially collective products, and individual agents rarely make a difference to their occurrence".[34] Consequently, where large (numbers of) countries choose to purchase hESC-derived therapies, it may make little or no actual difference to the amount of resulting embryo destruction if other (small) countries such as Fantasia purchase such therapies. This is perhaps Robertson's point, which we encountered earlier.[35] The actions of particular countries, and the encouragement that this would represent, may not be necessary for the commission of the wrong; the wrong would nonetheless occur on a widespread basis. It might be the case that the therapies would be produced regardless of Fantasia's stance since the demand created by other countries would be sufficient for research and production to take place. Accordingly, Fantasia's decision to import hESC-derived therapies would not be necessary in order for embryo destruction to continue.[i] As a result, it is difficult to tease out wherein the complicity lies if the moral wrong would take place anyway.

Earlier, in Section 2 of this chapter, we saw that where actors step into the causal chain of the moral wrong, they cannot claim that they are not complicit in it, even where the wrong would have taken place anyway. In the context of hESC research, I illustrated the point with the example of the pipette supplier; whether or not supplying the pipette makes no difference to whether or not that particular piece of research occurs, by supplying the pipette, a causal connection is made. It might be tempting to apply a similar line of reasoning to countries which import hESC-derived therapies. However, countries such as Fantasia whose decision regarding import does not affect the incidence of the wrong are not stepping into the causal chain in the same way that the pipette supplier does. They are in fact not making any causal contribution at all. It is not the case that the encouragement (or lack thereof) that Fantasia gives to ongoing embryo destruction through the import of therapies acts as a substitute for the self-same encouragement that would be provided by another country. With the arms dealer or the pipette supplier, their actions *do* prevent other arms dealers or pipette suppliers from acting in *those particular* circumstances at that particular time. This is to be contrasted with Fantasia, since its decisions or actions regarding import do not prevent other countries from also acting to import hESC-derived therapies.[j]

---

[i] In certain instances where the (potential) demand created by an individual country is relatively low, adopting a position which is pro-import might not even be sufficient for the research and development of hESC-derived therapies where other countries are not creating the demand.

[j] Notwithstanding that there might be a limited supply of hESC-derived therapies, in which case Fantasia importing them could prevent other countries from getting those same therapies. However, given that extra demand would probably be created, this would simply provide the encouragement to produce more therapies.

In this manner, if one contends that some sort of causal link needs to be maintained for charges of complicity to stand, it might be difficult to view the actions (or non-actions) of particular countries as constituting complicity. If, as I suggested above, in certain instances a country's position would not make a causal contribution to the purported wrong in any of these ways and, therefore, make no actual difference to the existence of the wrong in the world, that country need not be concerned about complicity in embryo destruction. It is, of course, the case that "assistance and encouragement often do make all the difference between commission and non-commission of the principal wrong"[36]; albeit, it might be tricky to determine. Be this as it may, this observation does not advance the argument for cases where the wrong truly would have occurred regardless of the encouragement offered through the actions of countries such as Fantasia. This is thus a problem of collective action where the actions of small parts of the collective are not necessary and do not by themselves affect outcomes. Yet an analysis which individuates out different actors fails to account for the cumulative and (more powerful causative) effect of the collective. While the encouragement given by Fantasia is not enough on its own to make a difference to the commission or non-commission of the putative wrong, it is likely that in concert with other countries it would be. Even if those other countries are also small and individually would not have a causative effect, together they make all the difference. There is, therefore, no point in disaggregating their actions, as to do so would be to conceal the sometimes complex nature of causation and hence moral culpability.

Consequently, we can see that the picture is complex, and that the demand created by individual countries for hESC-derived therapies would just be one part of a bigger whole. In this respect, Gardner maintains that in certain cases the principal wrongdoer is a collective, whereas the individuals which make up the collective can be seen as being accomplices.[37] This observation adds another layer of complexity to any analysis of the moral complicity in embryo destruction that is being discussed here. The reason is this: Earlier in the paper, I argued, contra Green, that the proximate cause of embryo destruction is hESC research. However, the research enterprise is in itself a collective, with each part being "just a ting cog in the huge machine of wrongdoing".[38] Some elements of the research machinery are more strongly connected to the wrongs done; for example, the hESC researchers who are the proximate cause of embryo destruction and are directly implicated in the causal chain. Yet when Fantasia permits the import of hESC-derived therapies, it does not act so as to directly encourage the individual researcher to destroy embryos, rather it acts such that it encourages the research enterprise as a whole. Fantasia acting in concert with other countries forms a collective which encourages another collective, the research enterprise, to commit the principal wrong. If this is correct, it might be that it is one or more steps down the chain of causation and thus one or more steps removed from being implicated in any sort of complicity. This, in effect, may weaken the connection that individual countries have to the principal wrong, yet it is difficult to see it completely removing the connection to the putative wrong of embryo destruction.

While the encouragement given by each individual beneficiary is not necessary, together it is sufficient to create demand for hESC-derived therapies and, hence, embryo destruction to take place. We might, therefore, think that each country ought to be held responsible for their own actions, including their contributory actions. We have just seen that the morality of those actions cannot be ameliorated *even if* the wrong would have occurred despite their own contribution. This is because, in concert with each other, individual countries make a difference by adding to the power of the collective. In such a situation, *even if*, despite the actions of individual countries, there would still be the same amount of the moral wrong in the world, countries which import hESC-derived therapies cannot claim 'clean hands' with respect to embryo destruction. This includes countries such as Fantasia which have taken all other reasonable legislative measures against embryo destruction. This means that countries which ban hESC research cannot import and use hESC-derived therapies without being implicated in the alleged moral wrong of embryo destruction, although the exact strength of the implication may be up for debate.

Given the arguments presented thus far in this chapter, complicity in the case of hESC research can be seen as functioning as a normative argument for a country's own internal moral regulation. Therefore, countries which ban hESC research on moral grounds ought to choose not to be complicit in the purported moral wrongs that lie therein. The morally conscientious and consistent State would not import hESC-derived therapies for use within its borders. However, such a situation is unlikely to arise where the State comes under pressure from the electorate to provide those therapies for its citizens. In such a situation, those in other countries who have worked to produce hESC-based therapies might feel aggrieved that a State which actively tried to inhibit scientific endeavour in this arena could stand to benefit. If this is the case, might there be an argument which would support a boycott in relation to those countries which ban the research on behalf of those who produce such therapies?

4. FAIRNESS AND RECIPROCITY IN (STEM CELL) RESEARCH

Those who do not believe that hESC research is morally problematic will not be troubled by concerns of complicity. After all, for them, no moral wrong has been committed and, therefore, they are not complicit in moral wrongdoing. They might, however, be concerned about the fairness of allowing those who actively try to prevent hESC research to benefit from the fruits of such research. Where countries such as Fantasia allow the import of hESC-derived therapies, one might argue that they are free-riding on the research efforts and resources of countries with more permissive stem cell regulations. One might argue that a country which has banned hESC research has not contributed to that research and therefore, should not be permitted to benefit from it. Benefiting from the fruits of other countries' research labour is not a problem because countries like Fantasia merely stand on the side-lines watching with moral condemnation. Rather, they take active steps to prevent such research taking place within their borders by enacting legislation and putting in place restrictive regulations. However, claims of fairness and free-riding across different jurisdictions and research environments might be more difficult to sustain than would first appear.

At first glance, it does seem that some sort of principle of fairness is contravened if those who not do not contribute to (and indeed actively try to hinder) certain scientific endeavours are subsequently allowed to become the beneficiaries of the hard work of other countries. The unfairness could be said to lie in the fact that there is a lack of reciprocity in such arrangements.[39] However, a practical look at the way that scientific and medical research operate on a global level shows us that it is not as simple as saying that Country A did not permit hESC research within its borders while Country B did, therefore Country A should be precluded from the benefits of any hESC-derived therapies. Different countries have different research priorities and, as such, conduct different research (even within the same broad areas). These diverse strands of research produce an equally diverse range of results. The results generated could, as such, be seen as contributing to the body of research knowledge as a whole. The research done in one country could thus aid further related research in another country.

In the context of stem cells, none of the various strands of research [adult, embryonic, foetal, somatic cell nuclear transfer (SCNT), or induced pluripotent stem cells (iPSC's)] is progressing in isolation of the others. For example, a research group working on adult stem cells in one country might make a discovery regarding basic cell biology or gene function which is then used by a group elsewhere that is conducting hESC research. In particular, those who oppose hESC research are touting research on iPSCs as being free of the moral taint that they believe accompanies research involving embryo destruction. However, as Katrien Devolder notes in a recent paper, "the fields of hESC and iPSC research progress in parallel and mutually support one another".[40] Specifically in relation to iPSC research, this is because hESCs are the analogue upon which iPSCs are modelled; the functionality which the iPSC researcher is after is based on the functionality of hESCs. Researchers, therefore, need to know a lot more about hESCs in order to model the iPSCs. The result of this is that rather than iPSC research reducing the need for hESC research, the likely outcome is that it will encourage such research.[41] There are two consequences of this. The first is that iPSC researchers (and the countries which permit their research) cannot claim clean hands in relation to the moral wrong of embryo destruction. The second is that any claims of unfairness and free-riding with regard to hESC-derived therapies might not hold up in the research context where developments are interrelated and interdependent. Interactions between research groups, and the transfer of knowledge that occurs between them, in effect would make it extremely difficult to determine a particular country's contribution to either the scientific effort as a whole or to stem cell science more specifically. As such, where it is reasonably thought that the benefits and burdens of this type of research (or indeed research in general) are fairly disseminated amongst the various countries or stakeholders, it would be difficult to try and stop particular countries from benefiting from hESC-derived therapies once they are available.

5. CONCLUSION

In this paper, I noted that in some countries, hESC research is considered to be morally acceptable and is therefore carried out, whereas in other countries it is either banned or substantially restricted. It is possible that such research will lead to therapies for a range of diseases, and it is here that the question central to this chapter arises: Should countries which ban hESC research be precluded from benefiting from hESC-derived therapies? Initially, I argued contra Robertson and Green that countries and their citizens can be seen as morally complicit in the purported wrong of the embryo consequent on hESC research. They do not commit the moral wrong themselves, but they do act with the effect that there is an incentive for the moral wrong to continue in the future. Nonetheless, I did note that, in trying to determine whether or not individual countries ought to be thought of as complicit, we run into a collective action problem. It can be difficult, if not impossible, to determine a particular country's causal contribution to the purported wrong. In this respect, I argued that individuating the different actors in an attempt to identify this contribution might in fact blind us to the cumulative effect that each country has in adding to the power of the collective. Consequent on this, countries which ban hESC research cannot import and use hESC-derived therapies without being implicated in the putative moral wrong of embryo destruction. As such, the morally conscientious and consistent State would not import hESC-derived therapies for use within its borders. I acknowledged, however, that such a situation is unlikely to arise where the State comes under pressure from the electorate to provide those therapies for its citizens. I therefore briefly examined the argument in virtue of fairness and free-riding to see if there might be a case for those who produce such therapies to boycott those who ban the research they are based upon. However, I argued that in the globalised research environment and economy, it would be difficult to dissect out each particular country's contribution to the research endeavour, both with regards to stem cell research specifically and research more widely.

REFERENCES

1. Gardner J. (2007) Complicity and Causality. *Criminal Law and Philosophy* **1:** 127–41, p. 127.
2. For further details, see http://www.clinicaltrials.gov/ct2/show/NCT01068951 (Accessed 16 June 2011).
3. Secrest AM. (2010) All-Cause Mortality Trends in a Large Population-Based Cohort With Long-Standing Childhood-Onset Type 1 Diabetes. *Diabetes Care* **33:** 2573–2579.
4. See http://diabetes.niddk.nih.gov/dm/pubs/statistics/#Types (Accessed 16 June 2011).
5. See the EuroStemCell website for more information — http://www.eurostemcell.org/faq/56 (Accessed 16 June 2011) — and the StemGen website, which contains a stem cell world map showing the regulatory approaches of different countries to human embryonic stem cell research — http://www.stemgen.org/mapworld.cfm (Accessed 16 June 2011).
6. For a discussion of possible rationales behind the seemingly morally inconsistent German law in this area, see Takala T & Häyry M. (2007) Benefiting from Past Wrongdoing, Human Embryonic Stem Cell Lines, and the Fragility of the German Legal Position. *Bioethics* **21(3):** 150–159.
7. Gardner J. (2007) Complicity and Causality. *Criminal Law and Philosophy* **1:** 127–41, p. 141.
8. Kadish S. (1987) A Theory of Complicity. In: Gavison R (ed), *Issues in Contemporary Legal Philosophy — The Influence of H.L.A. Hart.* Clarendon Press, Oxford, pp. 287–303, p. 294.
9. Gardner J. (2007) Complicity and Causality. *Criminal Law and Philosophy* **1:** 127–41, p. 138.

10. Gardner J. (2004) Book review of Kutz, C., *Complicity, Ethics, and Law for a Collective Age.* Cambridge University Press, Cambridge. *Ethics* 2004: 827–30, p. 828. See also Gardner J. (2007) Complicity and Causality. *Criminal Law and Philosophy* **1:** 127–41, pp. 138–9.
11. Gardner J. (2004) Book review of Kutz C. *Complicity, Ethics, and Law for a Collective Age.* Cambridge University Press, Cambridge. *Ethics* 2004: 827–30, p. 828.
12. For example, see Kutz C. (2000) *Complicity, Ethics, and Law for a Collective Age.* Cambridge University Press, Cambridge, and Kutz C. (2007) Causeless Complicity. *Criminal Law and Philosophy* **1:** 289–305.
13. Robertson JA. (2004) Causative vs. Beneficial Complicity in the Embryonic Stem Cell Debate. *Connecticut Law Review* **36:** 1099–1113.
14. Green RM. (2002) Benefiting from 'Evil': An Incipient Moral Problem in Human Stem Cell Research. *Bioethics* **16(6):** 544–556.
15. Robertson JA. (2004) Causative vs. Beneficial Complicity in the Embryonic Stem Cell Debate. *Connecticut Law Review* **36:** 1099–1113, p. 1103.
16. Robertson JA. (2004) Causative vs. Beneficial Complicity in the Embryonic Stem Cell Debate. *Connecticut Law Review* **36:** 1099–1113, p. 1104.
17. Robertson JA. (2004) Causative vs. Beneficial Complicity in the Embryonic Stem Cell Debate. *Connecticut Law Review* **36:** 1099–1113, p. 1105.
18. Robertson JA. (2004) Causative vs. Beneficial Complicity in the Embryonic Stem Cell Debate. *Connecticut Law Review* **36:** 1099–1113, p. 1105.
19. See Chapter 6, Section 4.2.2 in this volume.
20. See Chapter 6, Section 4.3.2 in this volume.
21. Mertes H, Pennings G. (2009) Stem Cell Research Policies: Who's Afraid of Complicity. *Reproductive Biomedicine Online* **19(1):** 38–42, p. 40.
22. Robertson JA. (2004) Causative vs. Beneficial Complicity in the Embryonic Stem Cell Debate. *Connecticut Law Review* **36:** 1099–1113, p. 1106.
23. Robertson JA. (2004) Causative vs. Beneficial Complicity in the Embryonic Stem Cell Debate. *Connecticut Law Review* **36:** 1099–1113, p. 1106.
24. Green RM. (2002) Benefiting from 'Evil': An Incipient Moral Problem in Human Stem Cell Research. *Bioethics* **16(6):** 544–556, p. 548.
25. Green RM. (2002) Benefiting from 'Evil': An Incipient Moral Problem in Human Stem Cell Research. *Bioethics* **16(6):** 544–556, pp. 548–50.
26. Green RM. (2002) Benefiting from 'Evil': An Incipient Moral Problem in Human Stem Cell Research. *Bioethics* **16(6):** 544–556, p. 553.
27. Green RM. (2002) Benefiting from 'Evil': An Incipient Moral Problem in Human Stem Cell Research. *Bioethics* **16(6):** 544–556, p. 554.

28. Green RM. (2002) Benefiting from 'Evil': An Incipient Moral Problem in Human Stem Cell Research. *Bioethics* **16(6):** 544–556, p. 555.
29. For a general overview of causation including the 'but-for' test, see Honoré AM. (2010) Causation in the Law. In: Zalta EN (ed.), *The Stanford Encyclopedia of Philosophy (Winter 2010 Edition).* Available at http://plato.stanford.edu/archives/win2010/entries/causation-law/ (Accessed 16 June 2011). See also Menzies P. (2009) Counterfactual Theories of Causation. In: Zalta EN (ed), *The Stanford Encyclopedia of Philosophy (Fall 2009 Edition).* Available at http://plato.stanford.edu/archives/fall2009/entries/causation-counterfactual/ (Accessed 16 June 2011).
30. Green RM. (2002) Benefiting from 'Evil': An Incipient Moral Problem in Human Stem Cell Research. *Bioethics* **16(6):** 544–556, p. 555.
31. See Gardner J. (2007) Complicity and Causality. *Criminal Law and Philosophy* **1:** 127–41, pp. 134–5. For a discussion and critique of the concept of a *novus actus interveniens* and of issues relating to proximate causation, see also Hart HLA, Honore´ AM. (1959, 1985) *Causation in the Law,* 2nd Edition. Clarendon Press, Oxford, pp. 95–108.
32. Gardner J. (2007) Complicity and Causality. *Criminal Law and Philosophy* **1:** 127–141, p. 141.
33. Devolder K, Harris J. (2005) Compromise and Moral Complicity in the Embryonic Stem Cell Debate. In: Athanassoulis N (ed.), *Philosophical Reflections on Medical Ethics.* Palgrave MacMillan, Hampshire, 2005, pp. 88–108, p. 94.
34. Kutz C. (2000) *Complicity, Ethics, and Law for a Collective Age.* Cambridge University Press, Cambridge, p. 113.
35. See Section 3.2 of this chapter and Robertson JA. (2004) Causative vs. Beneficial Complicity in the Embryonic Stem Cell Debate. *Connecticut Law Review* **36:** 1099–1113, p. 1106.
36. Gardner J. (2007) Complicity and Causality. *Criminal Law and Philosophy* **1:** 127–41, p. 138.
37. Gardner J. (2007) Complicity and Causality. *Criminal Law and Philosophy* **1:** 127–41, p. 140.
38. Gardner J. (2007) Complicity and Causality. *Criminal Law and Philosophy* **1:** 127–41, p. 140.
39. De Beaufort I, English V. (2000) Between Pragmatism and Principles: On the Morality of Using the Results of Research that a Country Considers Immoral. In: Gunning J. (ed.), *Assisted Conception: Research, Ethics, and Law.* Ashgate, Dartmouth, pp. 57–65, pp. 63–64.

40. Devolder K. (2010) Complicity in Stem Cell Research: The Case of Induced Pluripotent Stem Cells. *Human Reproduction* **25(9):** 2175–2180, p. 2178.
41. Devolder K. (2010) Complicity in Stem Cell Research: The Case of Induced Pluripotent Stem Cells. *Human Reproduction* **25(9):** 2175–2180, pp. 2177–2178.

8

Who Do You Call a Hypocrite? Stem Cells and Comparative Hypocritology

Søren Holm

Certain commentators claim that opponents of stem cell research are hypocritical, or, at least, will become hypocritical once stem cell derived treatments are developed. Such claims are based on identifying inconsistencies in arguments or actions, and from this imputing hypocrisy. This chapter examines whether these claims of hypocrisy stand up to scrutiny. First, it looks at the connections between claims of inconsistency and those of hypocrisy; second, it specifically explores whether benefitting from past wrongdoing is a form of hypocrisy; and finally, it examines the issue of dissenting individuals within the democratic state.

1. INTRODUCTION

The discussion about the ethics of human embryonic stem cell (hESC) research[a] in the United Kingdom and the United States has been

---

Centre for Social Ethics and Policy & Institute of Science, Ethics and Innovation, School of Law, University of Manchester, Oxford Rd., Manchester, M13 9PL, United Kingdom. Section for Medical Ethics, Faculty of Medicine, University of Oslo, P.O. box 1078, Blindern 0316, Oslo, Norway. Email: soren.holm@manchester.ac.uk

[a] In the following, 'stem cell research' will refer to 'human embryonic stem cell resesarch' and 'stem cell treatment' to 'treatment based on human embryonic stem cell research' unless specifically qualified.

characterised by a re-activation of old battle lines originally drawn during debates about abortion and assisted reproductive technologies. This has led to a situation where public debate has been polarised, and this polarisation has also influenced the academic debate within bioethics. One unfortunate side-effect of this is that there has been very little fruitful engagement between the two sides of the argument and very little willingness to try to take the arguments of the other side seriously.

In one corner of the debate, there has been a small cottage industry claiming that opponents of stem cell research are hypocritical, or will become hypocritical once treatments are developed based on stem cell research.[b,1–9] There are a variety of hypocrisy claims in the stem cell debate, but all share the core that some inconsistency is pointed out between the actions or arguments of a participant in the debate, and it is claimed that this inconsistency is a sign of, or amounts to hypocrisy. Some examples: Germany is claimed to be hypocritical because it does not allow embryonic stem cell research but is likely to allow the use of therapeutic stem cell products. George W. Bush is claimed to be hypocritical because he claimed to protect the lives of embryos but did not protect the lives of those who need stem cell therapies, or because he protected embryos while he waged war in Iraq.[c]

In this chapter, I will analyse these claims and very considerably extend the analysis I offered in 2006.[10] The chapter falls into three parts. The first part provides an analysis of the connections between different forms of ethical inconsistency and claims of hypocrisy. The second part then analyses the specific claims of hypocrisy that are connected to benefiting from past wrongdoing, and in passing notes the puzzling phenomenon that many who make this claim can make no sense of it within their own ethical theory. The third and final part will then engage in some comparative

[b] I apologise in advance for obliquely adding to that literature, but the temptation was too strong and, with Oscar Wilde "I can resist everything but temptation" (see note 9).

[c] Let me just for the record state that I, like any other good liberal, am happy to heap opprobrium onto George W. irrespective of whether it is completely justified or not in the specific instance.

hypocritology and analyse issues relating to dissenting persons within democratic states.

In order to situate the analysis, I will take as my point of departure some arguments put forward in his usual forthright style by my friend and colleague John Harris:

> It is perhaps not inappropriate to emphasize the bottom line, which is surely that if the use of human embryos for the creation of stem cells or for therapeutic cloning is unethical, as the French and the Germans, the Irish and the Maltese (through their laws and regulations) imply, then those nations will certainly not want to benefit from wrongdoing. One way of demonstrating this would of course be to agree in advance to forgo any benefits from such wrongful research and to deny their people the benefits of therapies that might be developed in this way. This is also the only consistent course for Americans who agree with George W. Bush.[11]

In the last part of this quote, we see a rapid and unargued move from "one way of demonstrating this" to "this is also the only consistent course". The purpose of this paper is partly to investigate whether this move is sound.

2. INCONSISTENCY AND HYPOCRISY

Claims of inconsistency and hypocrisy can attach either to actions or to arguments, and it is important to keep these two contexts apart.

Sometimes the inconsistency that underlies the hypocrisy claim is an inconsistency between two actions of a person, or between an action and some proposition the person has uttered (i.e. the standard notion of performative inconsistency). But actions cannot in themselves be inconsistent; they are only inconsistent under a certain description. In the case of actions, we therefore need to ask: What can we deduce concerning motives, reasons, commitments and justifications merely from the fact that we observe person A perform an action X of type T? Very little in the standard case. Apart from the well-known problem in providing generic descriptions and classifications of acts that do not in themselves already presuppose a specific evaluation of the action, we also have the problem that we do not know what kind of action A thinks he performed, e.g. he may not believe it to be of type T; why A performed X; whether A thinks X is justified; whether A knows all relevant facts about X and the consequences of X, etc.[d]

It is thus only if we ask A to justify himself that we can begin to derive normative commitments from his action, but by asking for justification we are already moving away from a context of action to a context of argument. There are cases where this epistemic uncertainty does not occur, namely the clear cases of performative inconsistency, where an actor has clearly specified that the action X is one that he will not perform, but such cases are rare.

In general, we have no reason to expect that actors can give a complete justification of their actions when first asked, and in most cases we may not even be justified in expecting them to give complete justifications on reflection. Even if particularism in ethics is false, this does not entail that most (or any) actors can actually derive a justification 'all the way down' to foundational ethical principles.

This means that observing that A performs two acts X and Y that are on some conception of ethics performatively inconsistent does not entail that A realised that they were inconsistent or should have realised that they were inconsistent, even if A holds the same ethical system as the person making the inconsistency judgement.[e] Deriving inconsistency claims merely from observed actions is therefore fraught with difficulty.

---

[d] None of these epistemic uncertainties preclude that we punish A for doing X if X is clearly an act of type T according to the legal system operating on A. This is partly because ignorance of the law is not disculpating, partly because the law does not take the normative systems of the actor into account but imposes its own classificatory scheme.

[e] Note that the inconsistency claim in itself requires that the observer has fully captured all morally relevant features of the contexts in which A acted.

In the case of observed political decision-making, we further have the problem that it is path dependent, patchy, often the result of compromise, and almost always what Sunstein calls an incompletely theorised agreement.[12] No one is willing or able to give a complete justification for the specific features of the decision, but everyone, or at least the majority are willing to live with it for the time being. It might even be the case that if we attempted to completely theorise the decision, i.e. justify it all the way down to basic principles, we could no longer reach agreement. This means that in politics, it is generally only for decisions that are strongly and overtly ideologically based that a full justification can even in principle be given.

In party political systems, the issue is further complicated because, whereas in principle every Member of Parliament votes according to his own conscience, in reality there are both questions of party discipline and questions of a necessary division of labour among Members of Parliament (MPs) that result in MPs not scrutinising all decisions in depth.[f] Not all MPs read every piece of legislation diligently (e.g. leaving that to the party's spokesperson on the particular issue), and not every MP makes up his own mind (e.g. leaving that to the party's spokesperson as well).

2.1 Consistency and Coherence

Let us move to the context of argument. Inconsistency is a fatal flaw in an argument; not only does an argument with inconsistent premises not lead to a sound conclusion, we also know that any conclusion can be derived from a set of inconsistent premises. It therefore seems to follow straightforwardly that anyone who is engaged in argument should aim to rectify any inconsistency in his argument as soon as it is pointed out to him. It is seemingly a pure demand of intellectual honesty. But is it really that simple?

---

[f] This again does not preclude that they are normatively responsible for their decisions to vote in a certain way. But it does mean that they are not necessarily 'argumentatively' responsible.

The first problem is that it is often controversial whether the inconsistencies that are pointed out in ethical arguments are really inconsistencies. The typical case where inconsistencies are claimed is not one of two open-minded interlocutors trying to find the truth as a collaborative enterprise, but of two long-standing opponents engaged in what is just the latest round of the match. So it is rarely the case of a single argument containing clearly inconsistent premises, but of two arguments or positions where it is claimed that argument a1 commits its presenter to something, whereas argument a2 commits him to something which is inconsistent with the commitment derived from a1. It is most often not two explicit premises that are inconsistent, but either putative enthymematic premises, some further 'explication' of a premise, or some commitment derivable if further 'plausible premises' are introduced. This 'explication' of the arguments is often not done in the most charitable way.

It is typical that the person who has put forward the two arguments will often deny the inconsistency by denying that the derived commitments actually follow from the argument, for instance by pointing out that they are only derivable within his opponents' framework, not within his own, or by pointing out that they overlook important subtleties in the argument.

A second problem is that even if there is real inconsistency, the two inconsistent propositions do not occur in isolation. They are always part of a larger set of propositions that will contain both ethical and non-ethical propositions, since ethical argument typically can only be complete if empirical premises are introduced (e.g. in the stem cell debate, premises concerning the biological potency of stem cell lines, or concerning the realistic therapeutic expectations of stem cell research).[g] It is clearly still important that all propositions in this larger set are consistent, but it is equally important that they are coherent, that they are connected by reciprocal justificatory ties. Whereas changing one of the inconsistent propositions or simply deleting it from the set will remove the inconsistency,

[g] We might call this set the 'set of ethically relevant propositions'.

it may not always improve coherence. Coherence can be negatively affected if the modification creates new inconsistencies in other parts of the set of propositions, or if important justificatory ties within the set are broken.

If new inconsistencies are introduced, they will then have to be resolved, thereby possibly introducing yet further inconsistencies elsewhere and thus leading to a potentially very drawn-out chain reaction of inconsistency resolution, at the end of which our set of propositions may be completely different from the one with which we started. Such a process is most likely to occur if the initial modification is radical in the sense of modifying a proposition in the set that is directly connected to many other propositions.[h]

Are we morally or intellectually required to perform such a root and branch modification in our set of ethically relevant propositions?[i]

Here, two considerations are relevant. First, whether we on reflection believe that our set of ethically relevant beliefs as a whole is more likely to be correct than the modification necessary to remove the first inconsistency. If we believe that our set as a whole is more likely to be correct, we should still try to remove the inconsistency, but we should try to resolve it in a way that preserves as much as possible of the original set. Second, whether the predictable outcome of the process is likely to be more consistent at the cost of being less nuanced. A simple way of removing inconsistency in a set of ethically relevant propositions is to derive them all from a single foundational principle, but it is unclear whether that kind of consistency is really preferable. It requires both a very strong version of universalism and it also in most cases requires us to give up very commonly held beliefs leading to intransitive ethical preferences (e.g. the

---

[h] This is of course often the very effect that is sought by those who point out the inconsistency. In the case of stem cell research, they do not only want a change in stem cell policy, they typically also want changes in abortion and end-of-life policy (this is true for both the conservative and the liberal 'inconsistency identifiers').

[i] This question is a close parallel to the question of what the correct response is to a disconfirming finding in the context of a Lakatosian research program.

proposition that the loss of one life cannot be outweighed by the alleviation of a billion mild headaches). As Larry Temkin and, more recently, Stuart Rachels and Alex Friedman have pointed out, the judgements/intuitions leading to intransitive preferences are so strong and well entrenched that any ethical system requiring us to give them up will generate numerous highly counterintuitive results.[13–15] Similar arguments can be mounted concerning incommensurable goods.

These considerations entail that sometimes inconsistencies should be dealt with by the least radical approach possible, and that sometimes it may actually be preferable to live with an inconsistency in order to preserve the central, more certain insights of the total set.

In addition to the general issues concerning the process of political decision making outlined above, a specific, seemingly inconsistency inducing feature of the regulatory process needs to be noted. We often institute legal rights to perform certain actions in circumstances where there is no belief that all actions of this kind are ethically acceptable or right, and we often prohibit certain actions without believing that all actions of this kind are unacceptable or wrong. Giving A a right to do X does therefore not entail a judgement that X is right or good but just that[j]: (1) X is not sufficiently wrong or bad to warrant prohibition, or (2) acceptable and unacceptable instances cannot be easily distinguished, or (3) that a prohibition of X could only be enforced through unacceptable privacy infringements, or (4) that prohibiting X would lead to a slippery slope of regulation. It therefore does not follow that the moral equivalence of two actions X and Y should necessarily lead to their legal equivalence.

As Wertheimer has pointed out in his work on exploitation, the idea that someone may have a right to do wrong initially sounds inconsistent, but is not inconsistent on further analysis.[16] Rights do not only protect right action.

---

[j] This is not intended to be an exhaustive list of instances where we may legally allow people to perform actions that are morally wrong. It is just pointing out some of the instances relevant to the stem cell debate.

Political decision making is also different from individual decision making in that it is common and acceptable in democracies that policies are reversed, not because new arguments have become available, or someone has changed their mind after deep reflection on the issue, but simply because the balance of power has shifted and there is no longer a majority for the old policy. This may happen even if the electorate has not changed opinion on the matter in question, but has changed opinion on other matters and voted accordingly and thereby accidentally also changed the balance in Parliament on the issue at hand.

2.2 Is it Hypocritical to be Inconsistent?

We are now in a better position to answer the question of whether it is hypocritical to be inconsistent. Let us first note the definitions of hypocrisy and its cognates in the Oxford English Dictionary:

> Hypocrisy:
> The assuming of a false appearance of virtue or goodness, with dissimulation of real character or inclinations, esp. in respect of religious life or beliefs; hence in general sense, dissimulation, pretence, sham.[17]
>
> Hypocritical:
> Of actions: Of the nature of, characterized by, hypocrisy.
> Of persons: Addicted to hypocrisy, having the character of hypocrites.[18]
>
> Hypocrite:
> One who falsely professes to be virtuously or religiously inclined; one who pretends to have feelings or beliefs of a higher order than his real ones; hence generally, a dissembler, pretender.[19]

From these definitions, it is clear that hypocrisy is a very specific form of public, intentional inconsistency. It is public and intentional because it is deliberately aimed at misleading one or more other persons about the hypocrite's true character and motives. There are other ways of being hypocritical than by being inconsistent, but one clear form of hypocrisy is where someone publicly announces two inconsistent propositions, and where one or both of these is put forward with hypocritical intent.

From the intentional nature of hypocrisy and the analysis of inconsistency above follow a number of interesting conclusions about the relationship between inconsistency and hypocrisy, even if we assume that the inconsistency in question on reflection would be accepted as a true inconsistency by all parties in the debate.

Someone putting forward one argument a1 at time t1 and a different argument a2 at a later time t2, where a1 and a2 are inconsistent, can only be a hypocrite if he has noted the inconsistency, or if the inconsistency is so glaring that anyone ought to have noted it.

Someone putting forward one argument a1 at time t1 and a different argument a2 at a later time t2, where a1 and a2 are inconsistent, is only a hypocrite if after considerations of coherence the inconsistency is one that should be resolved, and if he has hypocritical intention in uttering a1 or a2.

Someone putting forward one argument a1 at time t1 and a different argument a2 at a later time t2, where a1 and a2 are inconsistent, is only a hypocrite if he has not changed his mind in the interval t1–t2, and if he has hypocritical intention in uttering a1 or a2.

These conclusions also hold for individual policy makers. For collective policy makers (e.g. parliaments), or individual policy makers that are effectively collective (i.e. Ministers in systems where they have to rely on the will of Parliament, or on negotiation with other parties), a slightly different set of arguments follows.

A policy maker promulgating one regulation r1 at t1 and a different regulation r2 at t2, where r1 and r2 are inconsistent, is only hypocritical if the political situation has not changed between r1 and r2, and if it has hypocritical intention in promulgating r2 or hypocritical intention in the manner in which the relation between r1 and r2 is displayed. Not changing r1 after the inconsistency has been pointed out does make the policy maker necessarily hypocritical only if changing r1 is possible without significant expenditure of political capital and without introducing even worse inconsistencies in the total set of regulations.

From these interim conclusions, we can finally deduce that whereas many people in the stem cell debate (and in most bioethics debates in general) are inconsistent, only few are positively hypocritical.

3. HYPOCRISY AND BENEFITING FROM PAST WRONGDOING

A particular type of hypocrisy claim that is often made in the context of stem cell research is the claim that it is wrong to benefit from past wrongdoing and therefore wrong to benefit from stem cell treatments if you believe that stem cell research is unethical.[k] What are we to make of this claim? There are potentially many ways in which stem cell research can be unethical, but let us for the sake of argument simplify and focus on the claim that what is ethically wrong about stem cell research is that the derivation of stem cells involves the destruction of embryos.

The wrongdoing that a person or a society benefits from in using stem cell treatments is thus the prior, necessary destruction of embryos. Under what circumstances is benefiting from such wrongdoing in itself wrong?

Let us first note that there is a crucial difference between benefiting from my own wrongdoing and benefiting from the wrongdoing of others. If I benefit from my own wrongdoing, I cannot plausibly claim that I was not involved in, or responsible for that wrongdoing. If I benefit from the wrongdoing of others, there are many circumstances in which I can legitimately claim not to have been involved in or responsible for that wrongdoing. To use a standard example from the literature and mentioned in the previous two chapters, if I receive a transplanted kidney from a murder victim and I am not involved in the murder, I benefit from the wrongdoing of the murderer, but I am not responsible for that wrongdoing.

[k] See Devolder (Chapter 6) and Quigley (Chapter 7) in this volume for more on the argument regarding benefit from past wrongdoing, although the authors here do not make the claim of hypocrisy in relation to this.

A state that prohibits the derivation of stem cell lines from embryos within its own territory, that tries to get international conventions passed that prohibit such derivation elsewhere, and that works actively in international fora to abolish or limit funding for stem cell line derivation is arguably not responsible if other states decide to pursue stem cell line derivation and embryo destruction.

But would it nevertheless be wrong for such a state to benefit from the wrongdoing of other states? And would it be wrong for a person who had in a similar way done all he could do to resist stem cell derivation to benefit from the wrongdoing of others, including perhaps the wrongdoing of his own state?

In the context of stem cell research, Ronald Green provides a helpful typology of different ways in which an agent can be connected to and benefit from wrongdoing and a helpful analysis of when such benefiting is morally wrong. According to Green:

> ... benefiting from wrongdoing is prima facie morally wrong under any of three conditions: (1) when the wrongdoer is one's agent; (2) when acceptance of benefit directly encourages the repetition of the wrongful deed (even though no agency relationship is involved); and (3) when acceptance of a benefit legitimates a wrongful practice.[20]

The first of these conditions has obvious implications for the use of some stem cell treatments. If I direct someone to generate a stem cell line that I can use, I cannot claim that I am not responsible for the attendant destruction of embryos. This will, for instance, mean that a patient who seeks treatment with personalised stem cells created through somatic cell nuclear replacement techniques will be implicated in wrongdoing and will act in a way that is morally wrong.

But there are many potential stem cell treatments where the treatment has been developed long before the patient needs it, and where the generation of the stem cell line has happened long ago. The wrongdoer is, therefore, not the agent of the patient at the time of the wrongdoing.

Will the acceptance of benefit from such treatments directly encourage the repetition of the wrongful deed? This is essentially an empirical question. Will the action of buying stem cell treatment T from company C make it more likely that either C or other companies/researchers derive new stem cell lines and destroy embryos? Conceptually it can make it both more likely and less likely. If increasing sales cement C's position as the market leader, it may make it less likely; if it shows that there is a market niche to be exploited, it may make it more likely.[1] Whatever the answer to the general question is, it is extremely unlikely that the action of any single person will have any discernible effect on the risk that more stem cell lines will be derived; by extension, the same is true of small countries. Whatever the Maltese government decides to do when stem cell treatments become available will neither encourage nor discourage companies contemplating entering the stem cell market!

What then of legitimating a wrongful practice? The mere purchase and use of a stem cell treatment does not in itself legitimate stem cell research. No one in the debate denies that good may come of evil, what is denied is merely that the evil is justified by the good outcome. It is therefore perfectly possible to combine the use of a stem cell treatment with the denunciation of those who developed it, just as in the transplant case mentioned above, I can denounce the murderer and legitimately feel no gratitude towards him, while still being glad that I have received a kidney transplant.

In most cases, buying and using stem cell treatments does not therefore involve benefiting from past wrongdoing in a morally problematic way. This entails that it does not involve inconsistency, whether argumentative or performative, and that it *a fortiori* does not involve hypocrisy.

In this context, it is interesting but also somewhat puzzling that many of those who make these claims related to benefiting from past wrongdoing

---

[1] In this context, it is important to note that stem cell products containing cells are highly complex, and that it will be almost impossible to create cheap generic versions. Any company entering the market will have to bear the full development costs.

cannot make sense of them within their own ethical theory. Within a generally consequentialist framework, there can be nothing in principle wrong in benefiting from past wrongdoing, whether it is your own wrongdoing or the wrongdoing of someone else. Let us first note that a consequentialist can never have an absolute prohibition against benefiting from wrongdoing, since a consequentialist can never have an absolute prohibition against anything apart from not maximising good consequences. But more importantly, consequentialism is purely forward looking. The only right-making feature of actions is whether they produce good consequences and benefiting from wrongdoing is almost by definition a possible way of producing good consequences. For the consequentialist, the wrongs of the past are simply 'sunk costs' that should not influence our current decision making, except in so far as they affect the future consequences of our actions.

A consequentialist putting forward a hypocrisy claim relating to benefiting from past wrongdoing is therefore putting forward an argument that he believes is logically valid but unsound! The necessary major premise that 'it is wrong to benefit from past wrongdoing' is one he himself believes to be false. All the work put into explicating and refining this premise is not put in in order to develop a better argument, but purely to develop an argument that can better embarrass the opponent.

4. COMPARATIVE HYPOCRITOLOGY AND DEMOCRATIC POLITICS

In the paper by John Harris that I have quoted above, Harris goes on to discuss whether benefiting from wrongdoing could be acceptable if the wrongdoing is the result of democratic decision making. He writes:

> The suggestion that we act inconsistently or hypocritically if we accept the benefits of processes we believe to be wrong or even evil seems harsh and even perhaps simply wrong. Consider a society which debates the proposal to build a dam to provide increased water resources and hydro-electric power. Some people object to the project on ethical grounds... [...] Though their objections are certainly ethical, they lose the argument

> and the democratic process approves the building of the dam. Would they then be wrong, hypocritical or inconsistent to use the electricity and water provided by the dam? It seems not, but what's the difference this case and that of embryo destruction? There are many, but one set of differences seems telling. In the case of the dame, the evils are significant but they are the sorts of evils which, *inter alia* the democratic process is there to balance against other comparable evils... [...] However, there are some evils that are so great that they cannot simply be resolved by subjecting the decision as to whether they should be done or not to the democratic process. Murder is surely one of these. That is why those who object to embryo research and instrumental use because they regard the embryo as one of us can and should neither accept the practices, nor the benefits of those practices, if they truly believe that killing an embryo is killing a person with the same moral and political status as you and me. If they do accept the fruits of practices which are in their judgement the moral equivalent of murder, they are not only "scoundrels and substractors", but also hypocrites and murderers.[21]

But this is surely wrong, as a brief excursion into the wonderful discipline of comparative hypocritology will show.

There are several instances where democratic states make democratic decisions that involve either murdering or not murdering people. If a state decides to go to war, it *ipso facto* decides to kill people. And if going to war is not justified in the particular case, that amounts to intentional, unjustified killing, or what we usually call murder. The claim that engaging in this particular war will amount to murder will often have been made by the opponents of the war. Similarly, a state that by democratic means decides to keep the death penalty engages in judicially sanctioned murder. There will always be miscarriages of justice, and for the person who is executed unjustly it cannot matter that this is done with judicial approval. Seen from his perspective, he is murdered by the state, and that is surely the most correct description of what happens. In the case of both engaging in an unjust war and retaining the death penalty, it may turn out to be the case that net benefit flows from these decisions and that we all end up benefiting from murder. If the Iraq war had led to a stable, democratic regime in Iraq, everybody would have benefited from this, whether or not the war itself was justified.

Another example that is perhaps an even closer analogy to the stem cell area because it involves benefiting from unethical research is the case of the ethical vegan who is a vegan because he believes that it is seriously wrong to kill animals. He may have been convinced by Tom Regan that animals have rights, including the right not to be killed, and he tries to live his life in a way that does not involve him directly or indirectly in the killing of animals. He is also active in the animal rights movement trying to affect a change of policy in relation to animals. The ethical vegan does, however, live in a democratic state where democratic processes have led to legislation and regulation which not only allow but require animal research as part of the development process of all new pharmaceutical products and medical devices. Would the ethical vegan be a scoundrel, substractor, hypocrite and murderer (of animals) if he uses pharmaceutical products when he is ill?[m] And more importantly for the question of this section, how should the ethical vegan act as a citizen of a democratic state?

In all of these examples, good may eventually flow from acts and policies that are in themselves (by hypothesis) very bad and unjustifiable. Harris makes two claims concerning democratic decisions of this type. The first is that they cannot in principle be resolved by submitting them to a democratic process; the second that this entails that those who believe the decisions to be wrong should refuse to benefit from them, and that if they do accept benefits, they are hypocrites and murderers.

The first of these claims is both right and wrong, and therefore fertile ground for equivocation. It is clearly right as an ethical proposition. Ethical questions are not resolved by democratic processes but by argument.[n] Yet this is true of any ethical issue, not just very contentious ones. The building of the dam does not become ethically right by being democratically legitimated. As such, Harris' first claim is as clearly wrong as a

[m] On Harris' line of argument, the ethical vegan would presumably, on pain of inconsistency and hypocrisy either (1) have to renounce his veganism (or at least convert to vegetarianism purely on health grounds) or (2) suffer and die.

[n] Argument may of course be a part of the democratic process, but let us assume that we are here discussing something that is finally decided not by agreement or consensus but by a vote.

proposition concerning the life of citizens in democratic states. In a democratic state, all issues of policy are eventually subject to political decision making no matter how important, contentious or divisive they are. Democratic decision making is not only about balancing interests. This is simply because the democratic process is the last recourse when there are no other ways to resolve policy questions that have to be resolved. And when they have been decided in a legitimate way, they have been resolved for that society. Not forever, but until there is a sufficiently strong political force to get them changed or reversed. Citizens who disagree with the policies and think them profoundly wrong may have a duty to try to get them changed; they should not just meekly accept them. However, this a duty to try to get them changed in ways that are compatible with living in a democracy. The function of democracy is not only as a decision-making device. Perhaps equally important is its function as a way to resolve contentious issues peacefully. If Harris' thinks otherwise, he owes us an explanation of what kind of actions those who disagree can legitimately engage in.

We can of course move issues from the overtly political arena of voting in Parliament by instituting fundamental rights either in human rights conventions or in constitutions, and giving responsibility for interpreting these documents to the courts. But instituting a specific set of rights is in itself a decision made by political means, and legal interpretation of such rights in itself often a political act.

What about the second of Harris' claims? Is it correct, even if his first claim is incorrect? The discussion above of benefiting from past wrongdoing has dealt with this issue in great detail. The conclusion of that analysis was that for the individual, there is nothing inherently problematic in benefiting from wrongs that others have committed and that I have resisted to the best of my abilities, as long as my benefiting does not fall within one of the three categories defined by Green. But could it be claimed that I am in some way responsible for the actions of my government even though I have resisted them, merely by the fact that I am a citizen? And for how long will such responsibility last? Is Harris, for instance, responsible for all the many past wrongdoings of 'perfidious Albion' and obliged not to benefit from any of them?

I may be responsible for compensation to the victims of my government's wrongdoing simply by being a citizen and taxpayer, and this responsibility may last for a long time, but in the present case there are no victims to be compensated. The mere fact that the wrongdoing happened *here* cannot in itself be enough to make me responsible. It may well be almost true that all Germans are responsible for the actions of the Nazi government during the Second World War, but even in that case it can only be almost true. The descendants of those few who actively resisted cannot be responsible just by being German.

5. CONCLUSION

It is difficult to see that claims of hypocrisy can help the stem cell debate to progress in a constructive way. In one sense, they are just claims of inconsistency with added personal invective, since they not only impute inconsistency but personal malfeasance to one's opponents. This is unlikely to engender any desire in those opponents to take the counterarguments to their position seriously.

Claims of hypocrisy are therefore likely to harden the battle lines and contribute to the belief that there is no middle ground. But as Mertes and Pennings point out, those who do oppose stem cell research and do not want to be complicit in it often have a sophisticated analysis of moral complicity that allows distinctions to be made between different kinds of complicity (e.g. along the lines of Greene quoted above).[o,22] Claiming that opponents of stem cell research therefore have to choose categorically between absolute opposition to all aspects of stem cell research and treatment or the promotion of human health is simply a false dichotomy.

When we further add the complications in substantiating claims of hypocrisy analysed in this chapter, it seems safe to conclude that it is

[o] For another more nuanced view, see the paper by Takala and Häyry on the German legal position (see note 23).

perhaps preferable to attend to the "beam in our own [philosophical] eye" first before excoriating others for the mote in theirs.[24]

REFERENCES

1. Easton L. (2003) Man and Superman. *British Medical Journal* **326:** 1287–1290.
2. Harris J. (2003) Stem Cells, Sex, and Procreation. *Cambridge Quarterly of Healthcare Ethics* **12:** 353–371.
3. Kaczor C. (2005) *The Edge of Life: Human Dignity and Contemporary Bioethics.* Springer, Dordrecht.
4. Capron AM. (2001) Stem Cells: Ethics, Law and Politics. *Biotechnology Law Report* **20:** 678–699.
5. Devolder K. (2005) Advance Directives to Protect Embryos. *Journal of Medical Ethics* **31:** 497–498.
6. Devolder K. (2010) Complicity in Stem Cell Research: The Case of Induced Pluripotent Stem Cells. *Human Reproduction* **25(9):** 2175–2180.
7. Sperling S. (2004) Managing Potential Selves: Stem Cells, Immigrants, and German Identity. *Science and Public Policy* **31:** 139–149.
8. Harris J. (2008) Global Norms, Informed Consensus, and Hypocrisy in Bioethics. In: Green RM, Donovan A, Jauss SA (eds), *Global Bioethics, — Issues of Conscience for the Twenty-First Century.* Oxford University Press, Oxford, pp. 297–322.
9. Wilde, O. (1893) *Lady Windermere's Fan.* Forgotten Books Classic Reprint Series, 2010, p. 11.
10. Holm S. (2006) Are Countries that Ban Human Embryonic Stem Cell Research Hypocritical? *Regenerative Medicine* **1:** 357–359.
11. Harris J. (2008) Global Norms, Informed Consensus, and Hypocrisy in Bioethics. In: Green RM, Donovan A, Jauss SA (eds.), *Global Bioethics, — Issues of conscience for the twenty-first century.* Oxford University Press, Oxford, pp. 297–322, p. 320.
12. Sunstein CR. (1995). Incompletely Theorized Agreements. *Harvard Law Review* **108:** 1733–1772.
13. Temkin L. (1987) Intransitivity and the Mere Addition Paradox. *Philosophy and Public Affairs* **16:** 138–187.
14. Rachels S. (2005) Counterexamples to the Transitivity of 'Better Than'. *Recent Work on Intrinsic Value — Library of Ethics and Applied Philosophy* **17(Part IV):** 249–263.

15. Friedman A. (2009) Intransitive Ethics. *Journal of Moral Philosophy* **6:** 277–297.
16. Wertheimer A. (1999) *Exploitation.* Princeton University Press, Princeton.
17. Oxford English Dictionary Online. http://www.oed.com/viewdictionaryentry/Entry/90491 (Accessed 16 Jun).
18. Oxford English Dictionary Online. http://www.oed.com/viewdictionaryentry/Entry/90495 (Accessed 16 Jun).
19. Oxford English Dictionary Online. http://www.oed.com/viewdictionaryentry/Entry/90493 (Accessed 16 Jun).
20. Green RM. (2002) Benefiting from 'Evil': An Incipient Moral Problem in Human Stem Cell Research. *Bioethics* **16:** 544–556, p. 544.
21. Harris J. (2008) Global Norms, Informed Consensus, and Hypocrisy in Bioethics. In: RM Green, A Donovan, Jauss SA (eds.), *Global Bioethics, — Issues of Conscience for the Twenty-First Century.* Oxford University Press, Oxford, pp. 297–322, pp. 321–322 [footnotes omitted].
22. Mertes H, Pennings G. (2009) Stem Cell Research Policies: Who's afraid of Complicity? *Reproductive Biomedicine Online* **19(suppl. 1):** 38–42.
23. Takala T, Häyry M. Benefiting from Past Wrongdoing, Human Embryonic Stem Cell Lines, and the Fragility of the German Legal Position. *Bioethics* **21:** 150–159.
24. See Matthew **7:** 3–5.

9

Stem Cell Research and Same-Sex Reproduction

Thomas Douglas, Catherine Harding,
Hannah Bourne, and Julian Savulescu

Recent advances in stem cell research suggest that in the future it may be possible to create eggs and sperm from human stem cells through a process that we term *in vitro* gametogenesis (IVG). IVG would allow treatment of some currently untreatable forms of infertility. It may also allow same-sex couples to have genetically-related children. For example, cells taken from one man could potentially be used to create an egg, which could then be fertilised using naturally produced sperm from another man to create a genetically-related child with half of its DNA from each of the men. In this chapter, we consider whether this technology could justifiably be denied to same-sex couples if it were made available as a fertility treatment to different-sex couples. We argue that it could not.

1. INTRODUCTION

Consider the following hypothetical case:

(*Jack & Jill*) Jack and Jill present to a fertility specialist. The couple would like to have a genetically-related child but without assistance they are unable

Oxford Uehiro Centre for Practical Ethics, University of Oxford, Suite 8, Littlegate House St Ebbes Street, Oxford OX1 1PT. Email: thomas.douglas@philosophy.ox.ac.uk.

> to do so. Jack produces an abundance of healthy sperm. However, Jill is completely unable to produce eggs or bear children as she has no ovaries or uterus — they were removed as part of a cancer treatment. Fortunately, a new technology is available that can overcome the problem. The technology will allow the creation of eggs from Jill's somatic (body) cells through a process known as *in vitro* gametogenesis (IVG). Genetic material from one of Jill's somatic cells will be transferred into an enucleated oocyte (an egg that has had its nucleus removed) via somatic cell nuclear transfer, creating an embryo which is a clone of Jill. Once the embryo has reached the blastocyst stage or beyond, stem cells will be harvested and induced to mature into eggs. These will then be fertilised with Jack's sperm before being implanted into the uterus of a surrogate, who will gestate the pregnancy.

The treatment described in this case is not yet feasible, but it may become so. Major advances towards IVG have been made in the mouse. Both sperm-like[1–4] and egg-like[5–7] cells have been derived from mouse embryonic stem cells *in vitro*. One laboratory also reported the production of live offspring following fertilisation of natural mouse eggs with sperm-like cells derived from embryonic stem cells.[8] The technique was highly inefficient, with 210 oocytes used and 65 embryos implanted in order to create the 12 live animals that were born.[9] In addition, only seven of the mice had genetic material from both parents, and all died prematurely. However, the creation of live offspring following *in vitro* gametogenesis was a significant preliminary step. Subsequent work has demonstrated that functional sperm can be produced, partially *in vitro*, from mouse epiblast cells.[10] The epiblast cells are pluripotent stem cells harvested from the embryo at a slightly later stage (post-implantation) than traditional embryonic stem cells.[11] These cells were induced to form primordial germ cells (gamete precursors) *in vitro* before being transplanted into mouse testes to produce sperm, which were capable of fertilising eggs to produce healthy and fertile offspring.[12] This is the first proof of principle for partially *in vitro* gametogenesis in mice.

Progress in humans has been slower, but researchers have been able to derive cells expressing markers specific to mature germ cells from human embryonic stem cells.[13,14] Testing the reproductive functionality of these cells is challenging due to ethical constraints placed on human research.

However, further scientific advances combined with changes in regulation may enable the creation of functional gametes and embryos capable of normal development. The fertility treatment described in *Jack & Jill* may eventually become feasible.

Suppose it were already feasible. Should Jack and Jill be permitted to take advantage of the treatment? Some might be reluctant to permit the treatment on the grounds that it requires the harvesting of eggs from a third party (although the same technology could also enable the creation of a plentiful supply of eggs).[15] Others might object to the creation and destruction of a human embryo.[16] However, now suppose that these issues can be avoided; a somatic cell taken from Jill can instead be converted directly into a stem cell which can then be used to derive eggs. That is, one of Jill's own somatic cells, for example, a skin cell, could be used to produce a healthy supply of her own eggs. Though this possibility is even more speculative, it is not entirely fantastic. It has been shown that somatic (body) cells can be induced to become immature cells with the characteristics of stem cells, termed induced pluripotent stem cells (iPSCs).[17–23]

Given this amendment, we suspect that many would judge that Jack and Jill should be permitted to utilise IVG to produce a genetic child, at least provided that the treatment is reasonably safe, that Jack and Jill are prepared to pay its full cost, and that there are appropriate safeguards in place to prevent the exploitation of surrogates. It is true, of course, that Jack and Jill could instead adopt a child, or create a child with the help of sperm and egg donors and a surrogate. However, couples often have a preference for genetically-related children, and this preference is widely accepted as legitimate; many societies have been willing to make assisted reproduction technologies such as *in vitro* fertilisation and intra-cytoplasmic sperm injection (ICSI) available to couples who wish to have genetically-related children. In some cases, these technologies have even been publicly funded.

Now consider the following case:

> (*Hamish & Harry*) Hamish and Harry present to a fertility specialist. The gay couple would like to have a child genetically related to them,

> but without assistance they are unable to do so. Hamish is able to produce an abundance of normal sperm, however Harry is completely unable to produce eggs or bear children as, being male, he has no ovaries or uterus. Fortunately, a new technology is available that can overcome the problem. The technology, IVG, will allow the creation of eggs from Harry's somatic cells. The procedure will then be similar to that described in *Jack & Jill.*

As in *Jack & Jill*, the treatment described in *Hamish & Harry* may become technically possible. Recent research suggests that it may become possible to take somatic cells from a man, generate stem cells via either somatic cell nuclear transfer or iPSC technology, and then generate eggs from those stem cells.[24,25] These eggs could then be fertilised with another man's sperm before being implanted in the uterus of a surrogate. Some experts doubt whether this will become feasible in humans.[26] Moreover, at this point it seems that it will prove extremely difficult to create sperm from female cells due to the absence of the Y chromosome which holds a gene responsible for initiating sperm development. Nevertheless, it would be bold to rule out the possibility that same-sex reproduction may become feasible through further developments in stem cell research.

Suppose that IVG were already feasible, and safe, in same-sex couples. Should Hamish and Harry be permitted to take advantage of IVG? If Jack and Jill should be permitted to use the technology, then it is difficult to see how it could justifiably be denied to Hamish and Harry. After all, the situation of the two couples is very similar.

The intuitive case for making IVG available to Hamish and Harry can be strengthened by considering another case:

> (*Roberta & Rex*) Bob wants to have a sex change. He does so and becomes Roberta, who is legally recognized as a woman. Roberta (formerly Bob) then meets Rex, and they begin a relationship. After a few years, Roberta and Rex decide that they want to have a baby, but since Roberta has no ovaries, they cannot conceive. Fortunately, it is now possible to create eggs using Roberta's somatic cells via IVG.

So long as we allow sex change operations, there is at least a case for accepting the use of IVG by someone like Roberta: IVG could be regarded as merely providing Roberta with a *more effective* sex change. Not only does Roberta now have the physique of a woman, but she can also reproduce as a woman. Moreover, if we grant that IVG should be made available to Roberta, it seems doubtful that it could justifiably be denied to a same-sex couple. How could we deny the use of IVG to a man wishing to reproduce with his male partner as a man (Harry), while granting access to a biological man who wishes to reproduce as a woman (Roberta)? Surely it is an unreasonable imposition to expect the gay man to undergo a sex change operation in order to be able to reproduce with his partner.

Nevertheless, we suspect that many people would be reluctant to make IVG available to Hamish and Harry. In the United Kingdom, a group of MPs have already argued (unsuccessfully to date) for an existing prohibition on IVG as a fertility treatment to be lifted.[27,28] However, they accepted that the technology should be banned for same-sex couples, largely because they recognised that the prospect of same-sex reproduction was unlikely to be politically palatable.[29,30] Clearly, widespread resistance to same-sex reproduction is expected.

In this chapter, we consider whether the opportunity to have genetic children via IVG could justifiably be denied to same-sex couples while being provided to different sex couples. We consider four arguments for a difference in access. These hold respectively that:

- Same-sex couples have chosen to be infertile, infertile different-sex couples have not.
- For different-sex couples, the inability to have genetically-related children is a deviation from normal functioning, for same-sex couples, it is not.
- It is contrary to nature to create a child descended from same-sex parents.
- Children descended from same-sex couples would typically have lower well-being than other children (including those created via IVG from different-sex parents).

We will argue that none of these arguments is persuasive. Throughout, we will assume that IVG for same-sex couples ('same-sex IVG') would be similar to IVG for different-sex couples ('different-sex IVG') with regard to safety, the need for surrogacy and the need to create cloned embryos. We will also assume that the widespread availability of existing varieties of assisted reproduction, such as gamete donation, surrogacy and *in vitro* fertilisation to different-sex couples is justified. We will *not* assume that the availability of these techniques to same-sex couples is justified, though we believe that it is; presumably many who would object to same-sex IVG would object to *all* assisted reproduction for same-sex couples, though they might object much more strongly to same-sex IVG, since it alone would allow same-sex couples to have *genetically-related* children.

2. CHOICE

Hamish and Harry can be regarded as suffering from the same condition — infertility — as Jack and Jill. It might be argued that there is an important difference between the two cases: Hamish and Harry's infertility was chosen in a sense that Jack and Jill's was not. More generally, it could be argued that same-sex couples have chosen to be in a relationship which invariably involves an inability to have children, whereas infertile different sex couples typically have not. This difference might then be used to defend denying IVG to same sex-couples while making it available to different sex couples. Perhaps it could be said that same-sex couples have, by choosing to be infertile, waived any claim they might otherwise have had to IVG. A person who chose to form a sexual relationship with a non-human animal would not have a claim to fertility treatments allowing reproduction with the animal. Likewise, it could be argued, a person who chooses to be with a member of the same sex in one's own species has no claim to fertility treatments.

Importantly, this argument does not rely on the controversial claim that people can choose their sexual orientation. Rather, it relies on the weaker claim that, regardless of their sexual orientation, people can choose whether to form same-sex or different-sex relationships. However, even if we grant this claim, the argument faces a serious difficulty. Though homosexual people sometimes do have the choice to form relationships with people of the opposite sex (and thus, perhaps, form a fertile couple), making this choice would have significant personal costs. It would require overriding one's sexual inclinations. It is likely also to require forming a relationship with someone whom one does not love. It may require dishonesty about one's own sexual orientation and motivations. It is doubtful whether we should require individuals to make such costly choices in order to have genetically-related children with their partners.

Consider the following case:

> (*Lyle & Leila*) Lyle and Leila, present to a fertility specialist. The heterosexual couple would like to have a genetically-related child, but without assistance they are unable to do so. Lyle has a low sperm count, while Leila produces very few eggs. Either would be able to have a child naturally with another partner of normal fertility. But as a couple, their chances of conceiving naturally are negligible. Moreover, both were aware of these facts before they formed a couple. Fortunately, there is a technology — intra-cytoplasmic sperm injection (ICSI) — that may enable them to conceive.

Like Hamish and Harry, Lyle and Leila knew before forming a relationship that they could reproduce naturally if they chose different partners. Moreover, the psychological costs for them of choosing different partners would probably have been lower than for Hamish and Harry, since at least they would not have had to over-ride their sexual orientation. Nevertheless, few would object to the provision of ICSI in this case on the grounds that Lyle and Leila could have had children naturally if they had chosen different partners. The presence of this option seems to provide no or insufficient reason for refusing access to assisted reproduction. It is therefore difficult to see how it could provide a sufficient ground for refusal of IVG in the case of a same-sex couple.

3. TREATMENT VERSUS ENHANCEMENT

A second argument for prohibiting same-sex IVG holds that whereas for same-sex couples infertility is normal, for different-sex couples it is not. Thus, while different-sex IVG could be regarded as a treatment — an intervention intended to restore normal functioning — same-sex IVG must instead be viewed as an enhancement — an intervention which aims to augment functioning to a supranormal level. This difference could be important. Some regard the distinction between treatment and enhancement as morally significant, holding that engaging in enhancement is presumptively morally problematic, perhaps even morally wrong, whereas undergoing treatment is not.[31,32] If IVG for same-sex couples counts as an enhancement, then perhaps there would be good grounds for prohibiting it, even while permitting IVG treatments for different-sex couples.

There is, however, scope to question the classification of same-sex IVG as an enhancement. Though infertility might be the norm for same-sex couples, it is not obvious that this is the relevant standard of normalcy. Perhaps we should instead focus on what is normal for 'romantic couples — sexes not specified'. If we consider romantic couples as a single group, disregarding whether they are same-sex or different-sex couples, then it seems clear that fertility is the norm. And if this is the relevant standard for normalcy, we could regard the infertility endured by same-sex couples as a deviation from what is normal in precisely the same way that the infertility endured by, say, 'incompatible' couples like Lyle and Leila is a deviation from the norm. On this interpretation, same-sex IVG could qualify as a treatment.

There is also scope to question whether reproductive assistance for infertile different-sex couples always qualifies as a treatment rather than enhancement. Some would hold that treatments must be responses to diseases, disorders or disabilities. It is not clear that infertility need be due to any of those things, especially in cases where it is caused by incompatibility between partners rather than absolute infertility in one member of the pair. It may be, then, that both same-sex and different-sex IVG would, in some circumstances, be enhancements.

In any case, pinning an argument for prohibiting same-sex reproduction on the treatment–enhancement distinction seems a risky strategy to take, since the moral significance of the treatment–enhancement distinction is hotly disputed. Two of the current authors have argued against its significance elsewhere.[33,34] Moreover, even those who regard it as significant and believe that enhancements should generally be prohibited are typically still willing to entertain that there might be good reasons to make exceptions to this blanket prohibition, for example, where there are good reasons to provide a particular enhancement. We believe that there are good reasons to make same-sex IVG available even if it is an enhancement. Doing so would allow many same-sex couples to achieve a highly valued outcome (having genetically-related children) and would not directly harm others (a point to which we will return below).

4. UNNATURALNESS

New assisted reproduction technologies and other biomedical technologies that extend the boundaries of orthodox medicine are often alleged to be contrary to nature and therefore ethically problematic. No doubt some would object to same-sex IVG on similar grounds. In order to assess the suggestion that same-sex IVG would be unnatural, we should distinguish between some different senses of 'unnaturalness'. David Hume helpfully distinguished between three different concepts of nature; one which may be opposed to "miracles", one to "the rare and unusual", and one to "artifice".[35] This taxonomy suggests a similar approach to the concept of unnaturalness. We might equate unnaturalness with miraculousness, with rarity or unusualness, or with artificiality. Clearly, producing children via same-sex IVG need not be literally miraculous. However, it might well qualify as unnatural in each of the other senses. At least at first, same-sex IVG would certainly be unusual. It would also be artificial. To say that something is artificial is roughly to say that it involves human intervention, or perhaps certain types of complex or sophisticated human intervention (such as the use of *technology*). IVG is, arguably, a paradigmatic example of an artificial process.

Few would argue that same-sex IVG ought to be prohibited merely because it is unusual. However, some might maintain that it should be prohibited because it is artificial. This suggestion seems to have some implausible implications, however. Almost all medical therapies are artificial. Similarly, the works of art and the buildings we live in are plausibly artificial. Yet treating disease, making art and constructing buildings can be good. Moreover, for many of us, their artificiality does not seem to detract at all from their goodness; it is not as though medical treatment is objectionable insofar as it is artificial, but good insofar as it cures disease.

Perhaps it could be argued that though artificiality and unusualness are not *generally* problematic, they *are* problematic within the specific realm of human reproduction. Leon Kass argues that "… the severing of procreation from sex, love and intimacy is inherently dehumanizing, no matter how good the product",[36] while Margaret Sommerville holds that "the most fundamental human right of all is the right to be born from natural human origins that have not been tampered with by anyone else".[37] These claims could be interpreted as assertions of the view that artificiality has no place *in human reproduction.*

However, even this view has some implausible implications. Artificial 'tampering' with reproduction is now commonplace in developed nations, from the use of amniocentesis to detect abnormalities in the foetus, to the use of life saving medical experience and equipment to assist with the delivery of premature babies, to the use of surrogacy, *in vitro* fertilisation and artificial insemination by infertile couples. *In vitro* fertilisation was itself described as "against nature" at an early conference on its development in 1972.[38] Even the use of contraception to control fertility must be classified as artificial tampering with natural human reproduction. If Kass and Sommerville are right and artificial interference in human reproduction is objectionable, it seems that we must regard all of these practices as objectionable. But this is counter-intuitive (and is also inconsistent with our assumption that it is permissible to provide existing assisted reproduction services to different-sex couples). Even within the limited sphere of reproduction, it seems doubtful whether artificial interference is inherently objectionable.

To avoid this problem, a critic of same-sex IVG could appeal to a further concept of unnaturalness according to which to be unnatural is just to be contrary to some *natural moral law* — that is, roughly, a set of moral norms that is not man-made but is inherent in or derived from nature. On such a concept of unnaturalness, there is no question whether something unnatural is objectionable or wrong; being wrong is part of what it is to be unnatural. However, claiming that the creation of same-sex parented children is unnatural in this sense provides no argument or reason for thinking that it is wrong. It is simply a way of asserting that it is wrong. A further argument for why the putative natural moral law prohibits same-sex IVG would need to be provided.

One possible argument is suggested by the literature on sexual ethics and same-sex marriage, where a natural moral law has also been widely invoked. Drawing on Aristotelian and Thomist traditions, some natural law theorists opposed to homosexual acts and same sex marriage have sought to ground their arguments on the idea that reproduction is a basic human good, and, crucially, that (different-sex) marriages are a uniquely suitable setting within which to realise this good.[a] They have then argued that engaging in homosexual acts and recognising same-sex unions are objectionable because they tacitly deny the unique suitability of different-sex marriage for reproduction.[39–41] One can imagine an analogous argument being offered against the provision of same-sex IVG, which might also be thought to tacitly deny the privileged status of different-sex marriage as a setting for reproduction. We are not in a position to fully develop and critique this possible line of argument here. Instead, we will simply outline three major challenges that it would face.

First, the claim that different-sex marriages are a uniquely suitable setting for reproduction is a controversial philosophical or theological claim. Accepting such claims as a basis for legal prohibition would be inconsistent with liberalism, which requires that practices are prohibited only when harmful, or at least contrary to a moral norm that could not reasonably be rejected.

[a] Often left implicit is the further premise that assisted reproduction is not also a basic human good.

Second, at least in some liberal democracies, accepting the claim that different-sex marriages are a uniquely suitable setting for reproduction would be in tension with actual policy in other, related areas. Many states have, for example, sought to lower the legal hurdles faced by homosexuals seeking to form families, despite influential religious views which regard homosexuality as contrary to a natural moral law. Particularly relevant in the present context is that many jurisdictions have sought to enable same-sex couples to form families. For example, in the state of Victoria in Australia, the Charter of Human Rights and Responsibilities Act 2006 and subsequent legislation recognises same-sex parented families, and attempts have been made to improve the access of same-sex couples to biomedical interventions that enable them to found families. The Assisted Reproductive Technologies Act 2008 ensures that single women and same-sex couples have the same access to assisted reproduction as different-sex couples. In the Second Reading Speech, the Attorney-General stated that "persons seeking to undergo ART must not be discriminated against on the basis of [their] sexual orientation, marital status, race or religion".[42] The Act also introduces a presumption that if a woman in a lesbian relationship undergoes a procedure as a result of which she becomes pregnant, she is presumed to be the mother, and her female partner is presumed to be a legal parent of the child provided she was the woman's partner at the time she became pregnant and consented to the procedure.[43] It would be difficult to reconcile these reforms with a prohibition on same-sex IVG based on the idea that different-sex marriage is a uniquely suitable setting for reproduction.

Third, the possibility of IVG might in fact undermine the central premise in the natural law argument sketched above: The claim that different-sex marriage is a uniquely suitable context for reproduction. This premise has typically been grounded on the idea that, unlike same-sex unions, different-sex marriages are procreative in kind; they are unions of the sort that are capable of and "oriented to" realising one of the distinctive goods of marriage, namely, the creation and rearing of children.[44] In response to the possibility of same-sex couples creating children with the assistance of gamete donors, natural law theorists have fine-tuned this claim so that to be procreative in kind, a union must be capable of and oriented to creating children who "each can only have two parents and who are fittingly the primary responsibility (and object of devotion) of *those two parents*".[45] But if IVG were possible, same-sex couples could also be procreative in kind, even on this strict criterion; same-sex unions could produce children with only two parents who would then be raised by those same two parents. Natural law theorists opposed to endorsing same-sex unions would thus be deprived of their central basis for distinguishing between different-sex and same-sex unions. In response to this problem, opponents of homosexuality and same-sex marriage could seek to further fine-tune their account of what counts as a union that is of the procreative kind. They might hold that a couple must be capable of reproducing without technological help, or capable of reproducing through typical forms of sexual intercourse.[46] However, this will look suspiciously like an *ad hoc* amendment to the argument.

5. THE WELL-BEING OF THE CHILD

A more promising argument for denying IVG to same-sex but not to different-sex couples would appeal to the welfare of the resulting child. Considerations of well-being are, after all, well accepted as grounds for legal restrictions both in liberal theory and in actual liberal societies. *Child* well-being is arguably an especially important consideration in setting public policy, since children are vulnerable members of society, dependent on others for their welfare and well-being and without a voice in public debate. If it could be shown that children conceived via same-sex IVG are likely to have substantially lower well-being, on average, than other children — including children conceived through IVG by different-sex parents — then perhaps there would be grounds for selectively prohibiting the use of IVG by same-sex couples.

Note that prohibition of same-sex IVG could not be grounded on the claim that children born through this method would be *harmed* (in the standard sense of being made worse off than they would otherwise have been). Even if children conceived through same-sex IVG were significantly worse off than other children, they would not be harmed by their parents' decision to reproduce. After all, if the parents had not reproduced, those children would not have existed at all. Perhaps if the child's life was so horrible as to be not worth living, then we could say that the child had been harmed by being brought into existence.[47] But no one is suggesting that the lives of children born to same-sex genetic parents would be so bad. In standard cases they would be better off existing with same-sex genetic parents than not existing at all. So a prohibition on same-sex reproduction could not be justified on the basis that it is necessary to prevent harm to the future child.

Perhaps, however, concerns about the well-being of the resulting child could justify restrictions on access to IVG even though that child would not be harmed by the use of IVG. For example, perhaps having a child likely to have reduced well-being could be regarded as an impersonal harm — a harm 'from the point of view of the universe' — and could be regulated on this basis. But why should we expect children with same-sex genetic parents to have reduced well-being? And could a reduction in well-being really justify a prohibition on same-sex IVG?

5.1 Parenting Skills

One answer to the first question might be that same-sex couples would typically provide less effective parenting than different-sex couples. It might be argued that couples of different sexes typically have complementary parenting skills that same-sex couples lack. However, if this were true, then we would expect existing children raised by same-sex couples to be worse off, on average, than those raised by different sex couples. There is little evidence that this is the case. Empirical research has demonstrated that children raised by same-sex social parents fare no worse than other children on a range of standard criteria measuring relationships and development.[48] Studies conducted in a range of countries indicate that what is important is the quality of family relationships, the quality of the parenting, the level of support within the family, and the family's access to resources, not the structure of the family. A recent review of studies from 1978 to 2000 concluded that they "did not reveal evidence that children of lesbian mothers differed from other children on emotional adjustment, sexual preference, stigmatisation, gender role behaviour, behavioural adjustment, gender identity or cognitive functioning", and further that the studies were remarkably consistent in these findings.[49]

5.2 Discrimination and Disapproval

There are some more convincing reasons to suppose that children conceived through same-sex IVG might have lower well-being, on average, than other children (including children produced through IVG from different-sex parents). These children are, for example, likely to be victims of discrimination and social disapproval. Studies have shown that children raised by same-sex couples suffered bullying and discrimination because of the sexuality of their parents. In one study, half the children surveyed had experienced homophobia by the age of 10.[50] As the abovementioned studies suggest, this does not appear to have a significant impact on their overall well-being. However, children *genetically descended from* same-sex parents might be even more likely to suffer from discrimination and negative attitudes because of the 'all male' or 'all female' origins of their DNA. Additionally, it might be suggested that, to the extent that homosexuality is, or is taken to be, partially genetic in basis, children born of homosexual parents might be viewed as more disposed to homosexuality themselves. This could make them even more likely targets for discrimination or negative attitudes.

Having same-sex genetic parents may restrict one's well-being by making one a victim of discrimination or negative social attitudes. However it is difficult to see how this could justify prohibiting IVG for same-sex couples. The likelihood that a couple's child will be subjected to discrimination or negative attitudes is not normally regarded as providing good reasons to prevent reproduction. In some societies, children of black or mixed-race couples are subjected to discrimination, but we would not contemplate restricting their freedom to reproduce or access to assisted reproduction because of this. Not only would doing so significantly disadvantage already victimised groups, it would also be seen as implicitly condoning or encouraging the objectionable social attitudes in question.[51]

5.3 Disturbing Knowledge

Children conceived via same-sex IVG might also be disturbed by the mere knowledge of the circumstances of their conception, and this might reduce their well-being. One possibility is that the children would be disturbed by the technological nature of the IVG process. However, this seems unlikely to be a significant effect. Children conceived via IVF might also be disturbed by the technological means of *their* conception, yet there is no evidence of any difference in cognitive or emotional development between IVF children and children conceived naturally.[52] In any case, if children conceived through IVG from same-sex parents would be disturbed by the technological nature of the IVG process, surely children conceived through IVG from *different-sex* parents would be similarly disturbed, so it's not clear how this concern could justify different access to IVG for same-sex and different-sex couples.

Another possibility is that the children of same-sex genetic parents would be disturbed by knowledge of their genetic origins, that is, by the knowledge that their DNA comes from two persons of the same sex. It seems somewhat doubtful whether such knowledge would have a significant effect on well-being. Knowledge about one's origins is surely a minor determinant of well-being compared with other factors like health status, economic well-being, and the quality of one's relationships. Although it was disturbing to Victorians to discover that they were descended from a common ape ancestor, most enlightened people today have gotten over this, and it would, presumably, be more disturbing to discover that one was descended from an ape than from two male human beings.

For the sake of argument, let us grant that knowledge of their (currently) unusual genetic origins would nevertheless significantly restrict the well-being of the children conceived through same-sex IVG. We should also grant that the well-being of children conceived through *different-sex* IVG would not be restricted in the same way; their genetic origins would be similar to those of children conceived naturally (although if IVG involved the creation of cloned embryos, the genetic material would be passed down in a less direct way).[53]

Nevertheless, we question whether same-sex IVG should be prohibited merely because the resulting child might be disturbed by his or her genetic origins. Children descended from notorious criminals or tyrants might well be disturbed by their origins, but few would seek to prevent descendants of men like Pol Pot, Josef Mengele or Idi Amin from having children, for example, by restricting their access to assisted reproduction. Indeed, liberal societies have been willing to allow couples to reproduce (and access assisted reproduction) even when there are much more serious risk factors for reduced child well-being, for example, where the prospective parents are very poor, terminally ill, in a dysfunctional relationship, or carriers of a genetic disease. In most, if not all, of these cases we would regard the possible impediment to child well-being as insufficient to justify constraints on access to assisted reproduction. If this is the right approach, then it would clearly be unjustified to restrict access to same-sex IVG on the basis of a much milder impediment to child well-being — the concern that children would be disturbed by their unusual genetic origins. If there is any 'harm to the universe' involved in having children who might be disturbed in this way, it simply is not serious enough to justify withholding assisted reproduction.

6. CONCLUSIONS

Advances in stem cell research may make it possible for same sex couples to conceive genetic children via *in vitro* gametogenesis (IVG). IVG could also be used as a fertility treatment for some different-sex couples who are unable to reproduce naturally or using existing assisted reproduction technologies. We have examined four arguments for prohibiting same-sex IVG while allowing different-sex IVG.

The first argument attempted to justify different treatment for same-sex and infertile different-sex couples on the grounds that the former have chosen to be infertile whereas the latter have not. It failed because we can imagine different-sex couples (like Lyle and Leila) for whom infertility is also a choice, and to whom we would not deny assisted reproduction.

The second argument appealed to the treatment–enhancement distinction to justify distinguishing between same-sex and different-sex IVG. It claimed that different-sex IVG qualifies as a treatment whereas same-sex IVG would be an enhancement. We challenged this argument both by questioning the classification of different-sex and same-sex IVG as, respectively, treatment and enhancement, and by doubting the moral significance of the treatment–enhancement distinction. We also noted that there are good reasons to make IVG available to same-sex couples, and that these might be decisive, even if we grant both that it would qualify as an enhancement and that enhancements are generally morally problematic.

The third argument held that we should prohibit same-sex IVG because it is unnatural to create children with same-sex genetic parents. A problem was to identify a concept of unnaturalness according to which this is both true and morally important. Creating such children would certainly be unnatural in the senses of being unusual and artificial, but neither unusualness nor artificiality seems morally significant. Alternatively, it could be argued that same-sex IVG is unnatural (and thereby also immoral) in the sense of being contrary to a natural moral law. Perhaps this argument could be grounded on the claim that different-sex marriages are a uniquely suitable setting for reproduction. However, grounding a prohibition on such a controversial claim would be inconsistent both with liberalism, and with reforms which many actual liberal democracies have enacted. Moreover, this claim would be undermined by the possibility of IVG, which would call into question the thought that only different-sex unions are procreative in kind.

Finally, the fourth argument appealed to the claim that children conceived through same-sex IVG could be expected to have reduced well-being compared to children conceived through different-sex IVG, perhaps because they would be victims of discrimination or disapproval or would be disturbed by the knowledge of their genetic origins. However, even if this is true, it would not justify a prohibition on same-sex IVG. Prohibiting same-sex IVG on the grounds that the resulting children will be the victims of negative social attitudes could be viewed as implicitly supporting those attitudes and would be inconsistent with providing assisted reproduction to members of victimised racial minorities. Prohibiting same-sex IVG on the grounds that the resulting children will be disturbed by their genetic origins is inconsistent with providing assisted reproduction in a wide range of circumstances where the resulting children are at elevated risk of reduced well-being on certain dimension(s), for example, where the parents are in a non-ideal relationships.

We have been unable to identify any persuasive argument for distinguishing between same-sex and different-sex couples in regulating access to IVG. Unless such an argument can be provided, consistency requires that IVG should be made available both to different-sex and same-sex couples, or to neither. Granting same-sex and different-sex couples different liberties in respect of the use of this technology would be discrimination in the most fully pejorative sense of that term.

REFERENCES

1. Geijsen N, Horoschak M, Kim K, *et al.* (2004) Derivation of Embryonic Germ Cells and Male Gametes from Embryonic Stem Cells. *Nature* **427(6970):** 148–154.
2. Toyooka Y, Tsunekawa N, Akasu R, *et al.* (2003) Embryonic Stem Cells Can Form Germ Cells *In Vitro. Proceedings of the National Academy of Sciences* **100(20):** 11457–11462.
3. Nayernia K, Nolte J, Michelmann HW, *et al.* (2006) *In Vitro*-Differentiated Embryonic Stem Cells Give Rise to Male Gametes that Can Generate Offspring Mice. *Developmental Cell* **11(1):** 125–132.
4. Kerkis AA, Fonseca SASAS, Serafim RCRC, *et al.* (2007) *In Vitro* Differentiation of Male Mouse Embryonic Stem Cells into Both Presumptive Sperm Cells and Oocytes. *Cloning and Stem Cells* **9(4):** 535–548.
5. Kerkis AA, Fonseca SASAS, Serafim RCRC, *et al.* (2007) *In vitro* Differentiation of Male Mouse Embryonic Stem Cells into Both Presumptive Sperm Cells and Oocytes. *Cloning and Stem Cells* **9(4):** 535–548.
6. Hubner K, Fuhrmann G, Christenson LK, *et al.* (2003) Derivation of Oocytes from Mouse Embryonic Stem Cells. *Science* **300(5623):** 1251–1256.
7. Lacham-Kaplan O, Chy H, Trounson A. (2006) Testicular Cell Conditioned Medium Supports Differentiation of Embryonic Stem Cells into Ovarian Structures Containing Oocytes. *Stem Cells* **24(2):** 266–273.

8. Nayernia K, Nolte J, Michelmann HW, *et al.* (2006) *In vitro*-Differentiated Embryonic Stem Cells Give Rise to Male Gametes that Can Generate Offspring Mice. *Developmental Cell* **11(1):** 125–132.
9. Nayernia K, Nolte J, Michelmann HW, *et al.* (2006) *In vitro*-Differentiated Embryonic Stem Cells Give Rise to Male Gametes that Can Generate Offspring Mice. *Developmental Cell* **11(1):** 125–132.
10. Ohinata Y, Ohta H, Shigeta M, *et al.* (2009) A Signaling Principle for the Specification of the Germ Cell Lineage in Mice. *Cell* **137(3):** 571–584.
11. Tesar PJ, Chenoweth JG, Brook FA, *et al.* (2007) New Cell Lines from Mouse Epiblast Share Defining Features with Human Embryonic Stem Cells. *Nature* **448(7150):** 196–199.
12. Ohinata Y, Ohta H, Shigeta M, *et al.* (2009) A Signaling Principle for the Specification of the Germ Cell Lineage in Mice. *Cell* **137(3):** 571–584.
13. Chen HF, Kuo HC, Chien CL, *et al.* (2007) Derivation, Characterization, and Differentiation of Human Embryonic Stem Cells: Comparing Serum-containing Versus Serum-free Media and Evidence of Germ Cell Differentiation. *Human Reproduction* **22(2):** 567–577.
14. Clark AT, Bodnar MS, Fox M, *et al.* (2004) Spontaneous Differentiation of Germ Cells from Human Embryonic Stem Cells *In Vitro*. *Human Molecular Genetics* **13(7):** 727–739.
15. Testa G, Harris J. (2005) Ethics and Synthetic Gametes. *Bioethics* **19(2):** 146–166.
16. Mertes H, Pennings G. (2010) Ethical Aspects of the Use of Stem Cell Derived Gametes for Reproduction. *Health Care Analysis* **18(3):** 267–278.
17. Takahashi K, Yamanaka S. (2006) Induction of Pluripotent Stem Cells from Mouse Embryonic and Adult Fibroblast Cultures by Defined Factors. *Cell* **126(4):** 663–676.
18. Nakagawa M, Koyanagi M, Tanabe K, *et al.* (2008) Generation of Induced Pluripotent Stem Cells without Myc from Mouse and Human Fibroblasts. *Nature Biotechnology* **26(1):** 101–106.
19. Takahashi K, Tanabe K, Ohnuki M, *et al.* (2007) Induction of Pluripotent Stem Cells from Adult Human Fibroblasts by Defined Factors. *Cell* **131(5):** 861–872.
20. Park I, Zhao R, West JA, *et al.* (2008) Reprogramming of Human Somatic Cells to Pluripotency with Defined Factors. *Nature* **451(7175):** 141–146.
21. Drusenheimer N, Wulf G, Nolte J, *et al.* (2007) Putative Human Male Germ Cells from Bone Marrow Stem Cells. *Society of Reproduction & Fertility Supplement* **63:** 69–76.

22. Dyce PW, Wen L, Li J. (2006) *In Vitro* Germline Potential of Stem Cells Derived from Fetal Porcine Skin. *Nature Cell Biology* **8(4):** 384–390.
23. Yu J, Vodyanik MA, Smuga-Otto K, *et al.* (2007) Induced of Pluripotent Stem Cell Lines Derived from Human Somatic Cells. *Science* **318(5858):** 1917–1920.
24. Hubner K, Fuhrmann G, Christenson LK, *et al.* (2003) Derivation of Oocytes from Mouse Embryonic Stem Cells. *Science* **300(5623):** 1251–1256.
25. Lacham-Kaplan O, Chy H, Trounson A. (2006) Testicular Cell Conditioned Medium Supports Differentiation of Embryonic Stem Cells into Ovarian Structures Containing Oocytes. *Stem Cells* **24(2):** 266–273.
26. The Hinxton Group. (11 April 2008) *Consensus Statement: Science, Ethics and Policy Challenges of Pluripotent Stem Cell-derived Gametes.* Available at http://www.hinxtongroup.org/Consensus_HG08_FINAL.pdf (Accessed 16 June 2011).
27. Roberts M. (18 March 2008) UK MPs Reconsider Artificial Gamete Ban. *BioNews.* Available at http://www.bionews.org.uk/page_13333.asp (Accessed 16 June 2011)
28. Alghrani A. (2009) The Human Fertilisation and Embryology Act 2008: A Missed Opportunity? *Journal of Medical Ethics* **35(12):** 718–719.
29. Anon. (2 February 2008) Getting Ready for Same-sex Reproduction. *New Scientist.* Available at http://www.newscientist.com/article/mg19726413.000-editorial-getting-ready-for-samesex-reproduction.html (Accessed 16 June 2011).
30. Aldhous P. (2008) Are Male Eggs and Female Sperm on the Horizon? *New Scientist.* Available at http://www.newscientist.com/article/mg19726414.000-are-male-eggs-and-female-sperm-on-the-horizon.html (Accessed 16 June 2011).
31. Fukuyama F. (2002) *Our Posthuman Future: Consequences of the Biotechnology Revolution.* Profile Books, New York.
32. Sandel M. (2007) *The Case Against Perfection: Ethics in the Age of Genetic Engineering.* Harvard University Press, Cambridge, MA.
33. Douglas T. (2008) Moral Enhancement. *Journal of Applied Philosophy* **25(3):** 228–245.
34. Savulescu J. (2006) Genetic Interventions and the Ethics of Enhancement of Human Beings. In: B Steinbock (ed), *The Oxford Handbook of Bioethics.* Oxford University Press, Oxford, pp. 516–535.
35. Hume D. (1978) *A Treatise of Human Nature,* 2nd Edition. Clarendon Press, Oxford, pp. 473–475.
36. Kass L. (1997) The Wisdom of Repugnance. *The New Republic* **216(22)**.

37. Sommerville M. (2007) Children's Human Rights and Unlinking Child-parent Biological Bonds with Adoption, Same-sex marriage and New Reproductive Technologies. *Journal of Family Studies* **13(2):** 179–201, p. 198.
38. Jonsen A. (1998) *The Birth of Bioethics.* Oxford University Press, New York.
39. Finnis J. (1997) The Good of Marriage and the Morality of Sexual Relations: Some Philosophical and Historical Observations. *American Journal of Jurisprudence* **42:** 97–134.
40. Grisez G. (1993) *The Way of the Lord Jesus*, Vol. 2: Living a Christian Life. Franciscan Press, Quincy, IL.
41. Lee P. (2008) Marriage, Procreation, and Same-sex Unions. *The Monist* **91(3–4):** 422–438.
42. Victoria Parliamentary Debates, Assembly (10 September 2008) 3442 (Robert Hulls, Attorney General).
43. S.13 Assisted Reproductive Technologies Act 2008 (Victoria).
44. Lee P. (2008) Marriage, Procreation, and Same-sex Unions. *The Monist* **91(3–4):** 422–438, pp. 427 & 432.
45. Finnis J. (1997) The Good of Marriage and the Morality of Sexual Relations: Some Philosophical and Historical Observations. *American Journal of Jurisprudence* **42:** 97–134, p. 131 [emphasis in original].
46. Lee P. (2008) Marriage, Procreation, and Same-sex Unions. *The Monist* **91(3–4):** 422–438, p. 430.
47. Feinberg J. (1985) Comment: Wrongful Conception and the Right Not To Be Harmed. *Harvard Journal of Law and Public Policy* **8:** 57–77, pp. 70–72.
48. Millbank J. (2002) *Meet the Parents: A Review on the Research of Lesbian and Gay Families.* Available at http://www.qahc.org.au/files/shared/docs/meet_the_parents.pdf (Accessed 16 June 2011).
49. Anderssen N, Amlie C, Ytterøy EA. (2002) Outcomes for Children with Lesbian or Gay Parents: A Review of Studies from 1978 to 2000. *Scandinavian Journal of Psychology* **43(4):** 335–351.
50. Victorian Law Reform Commission. (2007) *Assisted Reproductive Technology & Adoption: Final Report* (Law Reform Commission, Melbourne), p. 33.
51. Little MO. (1998) Cosmetic Surgery, Suspect Norms, and the Ethics of Complicity. In: Parens E (ed), *Enhancing Human Traits: Ethical and Social Implications* (Georgetown University Press, Washington, DC), pp. 162–176.
52. van Balen F. (1998) Development of IVF Children. *Developmental Review* **18(1):** 30–46.
53. Mertes H, Penning G. (2010) Ethical Aspects of the Use of Stem Cell Derived Gametes for Reproduction. *Health Care Analysis* **18(3):** 267–278.

10

The Permissibility of Recruiting Patients with Spinal Cord Injury for Clinical Stem Cell Trials

Anna Pacholczyk and John Harris

Stem cell therapies offer the potential to bring a great benefit to many. However, research into their clinical applications is in its early stages and requires the participation of individuals in Phase I clinical trials. This raises particular ethical issues. For example, commentators have raised worries about the vulnerability of patients, especially those to be recruited in the acute post-injury stage. When stem cell trial participants are recruited shortly after the injury, there is a risk of therapeutic misconception and of depression, in addition to the problems associated with having only a limited time for deliberation. In this chapter, it is argued that neither lowered mood nor a willingness to accept a higher level of risk in the acute stage mean that participants are unable to give an informed and autonomous consent. Furthermore, it is argued that, even where patients are unable to consent, their participation, although not in their direct medical interest, might be permissible on the ground that it is in their interests broadly conceived.

Institute for Science, Ethics, & Innovation, School of Law, University of Manchester, Oxford Rd., Manchester, M13 9PL, United Kingdom. Email: john.harris@manchester.ac.uk

1. INTRODUCTION

Stem cell therapies offer the potential to provide new treatments for spinal cord injury (SCI). The significant effort in developing stem cell therapies culminated most recently in the initiation of clinical trials. Translational research in this rapidly growing field raises ethical issues about the permissibility of recruiting patients shortly (usually days) after the injury. Some scholars maintain that those who have suffered a spinal cord injury are a vulnerable population and argue that there are potential problems with the ability of this patient population to give informed consent to participation in research.[1] Illes *et al.*, focusing on the perspectives of patients, expressed concerns about the ability of patients, especially in the sub-acute stage, to make an informed decision regarding treatment, even though there might be potential benefits from early stem cell treatment.[2] The challenge, as noted by Taylor, is to find a way to understand how the patients weigh the alternative outcomes, so that with their probabilities, even given the unknowns, the choice is both ethical for the clinician to offer and not clearly inappropriate for a patient who decided to receive innovation therapy".[3]

Such concerns are worth addressing, in particular given the recent Food and Drug Administration-approved Phase 1 clinical trial conducted by the Geron Corporation. The trial aims to recruit participants 7 to 14 days post-injury and to use human embryonic stem cells (hESCs) derived from oligodendrocyte progenitor cells (GRNOPC1).[4] The primary goal of the trial is to assess the safety and tolerability of the treatment. This is "measured by the frequency and severity of adverse events within 1 year of GRNOPC1 injection that are related to GRNOPC1, the injection procedure used to administer GRNOPC1, and/or the concomitant immunosuppression administered."[5] Although a preliminary assessment of efficacy is stated as a secondary goal of this clinical trial, typically the aim of phase 1 clinical trials is to assess the safety and feasibility of new therapies and to determine dosages of drugs to be used in subsequent studies.[6] It might be hoped that a direct therapeutic benefit can be demonstrated, but this is *unlikely* in early trials. In such trials, often the design of early studies (lack of randomisation, small sample sizes) means that it is difficult to produce generalisable knowledge about efficacy.[a,7–10]

In this chapter, we look at general concerns regarding the vulnerability of patients in the immediate post-injury phase following an SCI. We investigate whether and in what ways patients in the acute stage of injury might be considered to be vulnerable. In particular, we look at whether a susceptibility to depression is a strong enough reason to exclude such patients from early phase stem cell trials. We also explore the general problem of therapeutic misconception in relation to phase I clinical trials, expressly addressing concerns about the possibility that vulnerable SCI patients might believe that they are receiving a novel, yet proven medical treatment.

2. VULNERABILITY

Vulnerability is a complex concept.[11] In the context of medical research, the use of this term differs from the everyday sense of the word. Here, the vulnerabilities that concern us are only those which call into question the ability of the patient to properly consent to research. In research involving stem cells as a therapy for SCI, vulnerability in the cognitive sense seems to be the most relevant type to consider. This is because the situation of the patient shortly after incurring the injury might affect their ability to give fully informed consent. This could, in part, be due to the time restricted circumstances in which they are required to make the decision to participate in the trial. It might not be feasible to provide all the information necessary for an informed decision given the time critical situation. As such, achieving the level of deliberation required for a fully autonomous and informed choice might not be possible. This in turn might increase the risk that participants in early stem cell trials might misunderstand the primary purpose

[a] Examples of other such clinical trials include ones studying the transplantation of autologous bone marrow derived cells (note 7), autologous olfactory mucosal grafts (note 8), and autologous olfactory ensheathing cells (note 9), although these studies typically recruited participants in later post-injury stages.

of the trials, despite an apparently appropriate disclosure of relevant facts and the purpose of the research (therapeutic misconception). These problems could be compounded if there is a high level of post-trauma depression or lowered mood, which compromises the ability of patients to make decisions. One might argue that this combination of factors renders those with acute SCI more vulnerable than they otherwise would be and, as such, that they ought not to participate in stem cell trials of this nature.

2.1 Vulnerable Means Depressed?

Let us address the last concern first. One reason why recruiting participants might be seen as problematic is that the high rates of post-injury depression might mean that patients are not able to make a fully autonomous decision about their involvement in a clinical trial. For example, Bretzner *et al.* point out that patients who have suffered an acute traumatic event experience stress, anxiety, fear, and depression 7 to 14 days post-injury (which is the recruitment period for the Geron trial). This occurs in degrees proportionate to the severity of injury[12] and, for that reason, the authors argue that these patients, in the weeks after sustaining a complete SCI, are a vulnerable population.[13]

Despite the prevalent opinion in the 1980s that depression after SCI was necessary and inevitable, there has been little evidence to support this assumption.[14] Rates of depression among patients shortly after SCI are estimated to be between 20% and 44%.[15] While this rate of post-injury depression is substantially higher than the 4% to 10% in the general population,[16] current research suggests that the occurrence of post-injury depression might be influenced by a number of factors. These include individual differences in personality and coping strategies employed to deal with new life circumstances.[17] Researchers also identified a number of risk factors, such as a history of substance abuse and a prior history of mental health problems.[18] Nonetheless, although many patients experience mild and transitory periods of depressed mood after the injury, this in itself does not imply that the depressed mood will reach a level of severity that would impair decision-making. Even if we assume that a diagnosis of depression would be a good reason to exclude patients from participating in research, an awareness of particular risk factors and early assessment could aid in the identification of those who are at a greater risk of developing symptoms of this severity.

There are two main reasons why some might think that patients with severe clinical depression should be excluded from participation in research. The first is that depression is correlated with poorer rehabilitation outcomes (both short and long-term), poor compliance with self-care needs, higher medical expenses, increased risk of suicide, and a higher rate of secondary complications.[19] This means that patients with a higher risk of depression might not be the ideal participants for clinical research. Pharmaceutical companies might not want to include those with a higher risk of depression since it might be easier to demonstrate safety and efficacy of a treatment if this population is excluded from early research (although, if early results are encouraging, one would expect that novel therapies would also be used for treating this patient population). Furthermore, patients with depression have a higher risk of suffering adverse effects from the procedures involved in research. It might be argued that in early-stage clinical trials, only the sub-population with the lowest risk of adverse effects should be included, unless inclusion of a higher-risk population is necessary to produce generalisable knowledge. On the other hand, participating in research can be seen to be in the patients' wider interests in the sense that methodologically robust and well-conducted medical research could be said to benefit everyone. As such, we perhaps have a good reason to support the involvement in early stem cell trials of those with a high risk of depression, even if that means that demonstrating the safety and efficacy of a treatment might be less straightforward if they were not involved.

The second main reason we might think that patients with severe clinical depression ought to be excluded from such trials is connected to concerns about the autonomy of those with symptoms of depression. It might be argued that we need to allow adequate time for depression to be diagnosed and the injury-associated psychological issues to be addressed (which could take months) in order to ensure that the decision to participate is made autonomously. However, the presence of symptoms, even when clinically severe, does not automatically mean that those patients are not capable of making decisions concerning serious medical procedures; this includes both medical treatment and participation in research. Moreover, excluding those with depression from beneficial research could be both unfair to them and counter-productive. Depression can be a normal response to certain adverse events. For that reason, it may well prompt the patient to think about, and perform a considered assessment of those events. We do not necessarily want to prevent this on the grounds that acute reactive depression itself may not be fully voluntary.

Some research suggests that a considerable percentage of patients continue to suffer from depression years after the injury.[20] That means that the importance of the symptoms of depression for our assessment of the vulnerability of patients is not solely a distinguishing characteristic of patients in the acute stage. If this is correct, what prompts worries about the autonomy of decisions in the acute stage might not in fact be the presence of the symptoms of depression.

3. INFORMED CONSENT

The most significant problems for participation in early stage stem cell trials might be those related to (1) the potential problems with conveying information about the risks (which is necessary for informed consent to be possible), and (2) the associated risk of therapeutic misconception given the nature and aims of phase I trials. Adequate information about the risks inherent in the trial is especially important given the fact that the transplanted hESCs (along with immunosuppression) might adversely affect the chance of spontaneous recovery in a subset of trial participants. It has been reported that 6% to 10% of patients initially assessed as having sustained a complete SCI show spontaneous functional improvement over time. This suggests that the initial lesion was in fact incomplete.[21] Patients sustaining injury would undergo a first standard operation aiming to decompress the spinal cord and stabilise the spine. When participating in research, such as the trial conducted by Geron, they would then undergo a second surgical intervention to transplant the cells within one to two weeks after the injury. In addition to the general risks inherent in cell transplantation (such as adverse reactions and the effects of immunosuppression), the transplantation surgery would expose patients to additional neurosurgical risks. This could include extending the lesion while transplanting the cells or introducing infection. If an adverse event occurs, consequent on the research, participation in a clinical trial could be seen as disadvantaging some patients in relation to the medical care that they would otherwise have received outside the trial.[22]

3.1 Therapeutic Misconception

Given the risks just mentioned, a number of commentators specifically highlight the importance of adequate information during the process of consent. Sub-acute complete SCI patients might be thought of as being at increased risk of therapeutic misconception; that is, conflating participation in a clinical trial with gaining access to a proven, albeit novel, medical treatment.[23] First-in-human clinical trials usually aim to produce socially valuable medical knowledge and often do not serve therapeutic functions.[24] The purpose of trials such as Geron's hESC-derived GRNOPC1 trial is to contribute to generalisable knowledge and not to deliver cutting-edge stem cell therapy. Similarly, the research is designed to optimise the gain in knowledge and not the needs or interests of patients; it is this which determines the interventions the trial participants receive.[25] Bretzner and colleagues suggest that it is unlikely that sub-acute complete SCI patients would understand the invitation to enrol in Geron's trial in these terms. The consent process would have to unambiguously explain that clinical benefits are highly unlikely, and that it is the value of the knowledge to be gained from the investigation, rather than the product's expected therapeutic effect, which is the purpose of this phase of clinical research.[26]

The possibility of there being a therapeutic misconception in relation to early stage stem cell trials highlights the importance of the function of the consent process. A robust consent process can be seen as a way of informing participants and of supporting the deliberative process. Illes *et al.* note this issue and propose that the consent process could be improved by using a staged consent process.[27] Although the systematic use of a staged consent model takes more time than obtaining consent in a single stage, it has been shown to contribute to more informed decision-making. Not signing the consent form immediately after the initial discussion was found to independently contribute to improved knowledge scores and led to more questions and active participant involvement in the consent process.[28] While the practicalities of implementing staged consent would need to be assessed where time for deliberation is short, a similar staged process could be used for clinical trials that aim to recruit patients 7 to 14 days post-injury. When changing the way the informed consent process is conducted, care would need to be taken to maximise the time available for deliberation within given constraints and to address participants' interests in taking into consideration the risks and benefits of the trial depending on their particular context, type of injury, and life circumstances.[29]

4. AUTONOMY, VULNERABILITY, AND LEGITIMATE REASONS

The time just after the injury might not be *optimal* for making decisions to participate in research. Yet, what we might be concerned about is not only maximising autonomy, as important a goal as that is, but also with determining whether patients in an acute stage can make a *sufficiently* autonomous choice (or, in other words, that the choices that they make fulfil the minimum required for the decision to be autonomous and informed). This raises the issue about how important the optimisation of autonomy is, as opposed to simply assuring that the minimum requirements for an autonomous and informed choice are met.

The value of maximising autonomy has to be weighed against the probability of therapeutic benefit to the patient and also against the risk of losing the possibility to pursue the most promising lines of research that could benefit many. A majority of studies suggest that the window for the transplantation of stem cells is 7 to 14 days post-injury, and only a small minority of research has pointed towards some benefits of this in the chronic stage post injury.[30]

Next, it is worth noting that our view of the ability of post-injury patients to make a sufficiently informed choice might be affected by our assessment of the reasonableness of their risk assessment. Bretzner and colleagues suggest that the risk of therapeutic misconception is especially high in patients shortly after SCI.[31] These patients suddenly and unexpectedly find themselves in dire circumstances and, because of this, they may fail to appreciate the disadvantages associated with trial participation. However, we need to distinguish concerns regarding the lack of appropriate information or awareness of the risks from those regarding the different weight that might be assigned to those risks in the decision-making process. Even where the patient is sufficiently informed about risks and benefits associated with participation in research, doubts might still be raised about the ability of the patient to consent.

Our view of the rationality of assigning weight to the particular known risks might influence whether we think that the patient is able to consent to research. For that reason, it might be worth having a closer look at how patients make decisions about participation in research. One of the interesting findings of Illes' study is that the type of injury, as well as the time from the traumatic event, influences patients' assessments of risk and the readiness to participate in stem cell research in a way that is perhaps not intuitive. For example, those within a chronic cervical group[b] were substantially more likely to be risk averse than those with a thoracic injury. In addition, those with chronic SCI were generally more reluctant to consider participation in research than those in a sub-acute stage. This suggests that the reasons for and against participation in stem cell trials may differ depending on the patient's life situation and their ability to adapt to new circumstances.

Such findings could be partly explained by different amounts of information about the potential for recovery and available therapies, along with

---

[b] The participants were divided into four groups depending on the type of injury (cervical, thoracic) and the time that had elapsed since the injury (sub-acute and chronic group).

the risks associated with them. Another probable explanation is that patients in the chronic stage have had time to adapt to new circumstances. This is expressed by one of the participants in Illes' study who said:

> I do worry about the newer injuries being first on the list because I know how that feels . . . all you want to do is walk again, whatever it takes, it doesn't matter. So that scares me a little bit.[32]

This suggests that the lack of sufficient information about the risks and benefits of trials might not be the only cause for the greater willingness to participate in research. Rather, given the fact that the patient did not adjust to the new, post-injury way of living, it seems reasonable that he or she is willing to accept a higher level of risk if associated even with a low probability of a substantial benefit. This seems understandable and perhaps rational even in the absence of therapeutic misconception. Yet, having a powerful reason to consent does not mean the consent is not genuine; if, given my priorities, this is such a good offer that I would be a fool to refuse, then I would be a fool to refuse. Given the circumstances, the small chance of benefit might understandably have a greater weight than at a later time. Absent other causes of autonomy being impaired, it is not at all obvious that the strong preference for walking again impairs one's autonomy to the extent that one is unable to consent to participation in stem cell trials. Consequently, as long as patients are well-informed, the acceptance of a high level of risk does not mean that the patient is not able to give a valid consent to research. It simply means that a small chance of benefit of a certain kind is a compelling reason to participate in research *in given life circumstances.*[c]

Consider an example that does not involve a medical context. Let us imagine Sam, a moderately gifted student who wants to become a professional dancer. He has a one-off chance to take part in a competition that, if won, will be a ticket to the world of professional dance. Sam decides to

---

[c] If, for whatever reason, the patient *is* judged to be unable to consent, then the decision of whether she should participate in stem cell trials could be assessed on the basis of their best interests.

leave his part-time job and does not attend his courses at university. Instead he focuses on preparation for the competition and spends all the money he has on this preparation, despite the fact that his chances of winning are slim. In doing so, he risks disastrous outcomes on the university exams. This might have implications for his future non-dancing career prospects. Although our opinion of whether it is worth the risk might differ from Sam's, we are likely to accept that Sam has compelling and understandable reasons to take the risk given what is important *to him* at that *particular* moment in time. Those reasons are compelling despite the fact that in several years' time, other enjoyments and goals may take greater importance and he may lead a fulfilling non-dancing life when his desire to be a professional dancer fades. As long as Sam is not lacking some crucial piece of information, has a reasonable idea about the low probability of winning, and accepts the risks, there is no reason to deny him the opportunity to take part in the competition on the basis of his desire to be a dancer being *too* strong.

The purpose of this example is to illustrate the point that a strong desire and associated acceptance of risks might be understandable. It also illustrates that as we adapt to new life circumstances, our priorities and preferences change. In fact, what seems to be difficult is the *process* of adaptation itself. Chronic cervical patients have reported that, even if a significant improvement in functioning was to be achieved as a result of participation in a stem cell trial, this change would require an enormous adjustment in their daily life. This would include going through the process of rehabilitation and redefining one's life plans and identity for a *second* time. As one of the patients put it, "[t]he transition was hard enough the first time; I don't want to go through it again".[33]

4.1 Only Fools Go First?

One reason why there might be doubts regarding the ability of patients in the acute stage post-SCI could be that 'choosing to go first', to participate in a phase I clinical trial, means that one is acting against one's best interests. Participation in trials of stem cell therapies might be judged to be choosing the option that is *worse* than the best available standard of care. However, is it really the case that participation in early clinical trials is against patients' interests? We do not think that this is necessarily so; there are other reasons to participate in a clinical trial, even if it might carry higher risks than an existing therapy that is considered to be the gold standard of care for SCI. The *potential* of stem cell therapies to one day relieve suffering experienced by the patient might provide a motivation to choose to participate in research. Even accepting the small probability of therapeutic benefit from this trial, participants might hope that their involvement will help to develop more effective cures than those currently available and that others, and perhaps even they themselves at a later time, will greatly benefit. A similar reason might provide a motivation to participate in research for those with chronic SCI who are less likely to receive a therapeutic benefit from research than patients in an acute stage post-injury.[34] If sufficiently informed, their willingness to participate might be thought to be a result not of vulnerability, but rather of the personal experience of suffering. Such suffering might provide an individual with compelling reasons to aid research that has resulted from that life experience.

However, acceptance of this line of reasoning places great responsibility on those conducting the trials to ensure that patients are not recruited lightly, and that the full reasons for attempting a trial, along with an honest assessment of its probable outcomes, are always fully explained. Any consent obtained for participation in such a trial should contain a full acknowledgement by the patient that they are significantly motivated by the opportunity to contribute to the public good and to the progress of science, as well as to improving the chances for future patients such as themselves.

4.2 Lack of Consent and the Duty to Participate in Research

It is important to remember that even where patients are *not* able to consent, this alone does not make their participation impermissible. One of the functions of informed consent to research is to protect participants from abuses such as those that motivated the development of the

Nuremberg Code. One of the ways of protecting those who cannot give informed consent is to require the proxy decision makers to act according to the person's best interests.[35] If we accept that early-stage clinical trials are unlikely to bring direct benefit to the patient, it might seem that participation in research is not in the patient's best interests. However, this interpretation of best interests considers only a narrow subset of patients interest and misses the powerful interest that persons have of living in a society in which beneficial research is pursued. Moreover, it is in our interests to be a serious moral agent, and by being taken by others to be a moral agent. If this is the case, acting on one's moral obligations is not contrary to one's interest, but rather is an integral part of one's interests; this is so even if acting on those obligations is not in one's interests narrowly conceived.

Elsewhere, one of the authors of this chapter has argued that, in certain specific circumstances, there is a moral obligation to participate in research.[36] This moral obligation can be justified, first, by our duty to help those in need and, second, by the 'free rider argument' that highlights the moral importance of contributing to beneficial social practices from which we ourselves benefit.[37,38] This obligation arises when a given research project is well-designed and likely to lead to important knowledge that will bring benefit to persons in the future, when the risks and burdens are not excessive, and is especially strong when there are no other potential participants for whom the costs and risks of participating are smaller.

5. CONCLUSION: ENHANCING THE DECISION-MAKING PROCESS UNDER TIME CONSTRAINTS

In this chapter, we did not aim to conduct a full empirical review of the kind of reasons for participation in research which people have. Nor did we attempt to argue that people who decide to participate in research are *always* fully autonomous, have a sufficiently good understanding of risk and benefit, or are free of therapeutic misconception. Rather, the aim was to explore some of the doubts expressed in the literature regarding the ability of patients in the acute stage of SCI to make sufficiently autonomous choices about participating in stem cell trials such as that being conducted by Geron. In addition, we reflected on the ethical permissibility of recruiting participants when their choices might be influenced by a strong desire to go back to the pre-injury level of functioning.

One could argue that the fact that there is not much time waiting 'for the dust to settle', or for deliberation, means that patients do not realise fully what their life is likely to be like after the injury. They may not realise either that there is the potential to lead a full and fulfilling life even with a limited motor function, or the difference in well-being that small changes in certain motor function might bring. The exact challenges that one is likely to encounter will not be obvious before the injured individual has experienced them for themselves.

Enhancing the autonomy of potential stem cell trial participants might be achieved by providing information on details of the kinds of futures they might expect both with and without receiving the stem cells. To the extent that it is possible, insights of those who have been in a similar situation, as well as empirical data about quality of life and the practical challenges that have to be faced after SCI will be valuable. Furthermore, the perspectives of those with chronic SCI are worth pursuing in order to gain information which would be relevant to the maximisation of the autonomy of any decisions made regarding participation. Empirical studies such as those conducted by Illes *et al.* can be of use in this respect. Studies examining patients' perspectives can help to inform us about the way patients make decisions and the level of their knowledge about available options. They can also be a source of information about the available alternatives (for example, the Illes study found that despite their interest, patients struggle to find reliable and comprehensible information about stem cell trials[39]). This can provide a needed aid in developing better ways of informing patients and designing consent procedures.

When consent to research is an option, a question that we are facing is whether the choice is a *sufficiently* informed and autonomous choice. Decision-making processes can always be improved; even fully autonomous agents can seek additional or better information, spend more time on deliberation of possible risks and outcomes, build complex forecasting models, and reflect on the decision-making process itself. However, although maximising autonomy in this sense is, generally speaking, a worthwhile goal, the constraints that are typically put on us (restrictions of time and resources) mean that at some point it is rational to stop thinking and start acting, even if that means running the risk of making sub-optimal decisions. The time constraints in the case of participating in clinical trials that recruit participants in the acute post-injury stage result in a window of opportunity that is narrow. There is a trade-off here of course, but in the words of Brutus in *Julius Caesar*:

> There is a tide in the affairs of men
> which taken at the flood, leads on to fortune;
> Omitted, all the voyage of their life
> Is bound in shallows and in miseries.[40]

Sometimes we cannot wait for a better state of the tide, and we would be crazy to postpone the decision if the result of postponement is well calculated to be misery. As Brutus said, "we must take the current when it serves or lose our ventures".[41] He got it wrong in his own case. He died a notoriously noble death, defending his beliefs and his honour, and trying to protect his fellow citizens from what he believed amounted to tyranny. He was surely right to take the chance, both for himself and for what he valued most. Likewise, it ought not to be considered wrong for patients in the acute post-injury stage after SCI to take the chance and to participate in early clinical stem cell trials.

REFERENCES

1. For example, see Illes J, Raimer JC, Kwon BK. (2011) Stem Cell Clinical Trials for Spinal Cord Injury: Readiness, Reluctance, Redefinition. *Stem Cell Review and Reports*, Online First, doi: 10.1007/S12015-011-9259-1; Taylor PL. (2010) Overseeing Innovative Therapy Without Mistaking it for Research: A Function Based Model Based on Old Truths, New Capacities and Lessons Learned from Stem Cells. *The Journal of Law, Medicine and Ethics*

38(2): 286–302; Robertson JA. (2010) Embryo Stem Cell Research: Ten Years of Controversy. *The Journal of Law, Medicine and Ethics* **38(2):** 191–203.

2. Illes J, Raimer JC, Kwon BK (2011) Stem Cell Clinical Trials for Spinal Cord Injury: Readiness, Reluctance, Redefinition. *Stem Cell Review and Reports.* Online first, doi: 10.1007/S12015-011-9259-1
3. Taylor PL. (2010) Overseeing Innovative Therapy Without Mistaking it for Research: A Function Based Model Based on Old Truths, New Capacities and Lessons Learned from Stem Cells. *The Journal of Law, Medicine and Ethics* **38(2):** 286–302. See also Robertson JA. (2010) Embryo Stem Cell Research: Ten Years of Controversy. *The Journal of Law, Medicine and Ethics* **38(2):** 191–203.
4. Geron. (2010) Phase 1 Safety Study of GRNOPC1 in Patients with Neurologically Complete, Subacute, Spinal Cord Injury. Available at http://clinicaltrials.gov/archive/NCT01217008 (Accessed 16 June 2011).
5. Geron. (2010) Phase 1 Safety Study of GRNOPC1 in Patients with Neurologically Complete, Subacute, Spinal Cord Injury. Available at http://clinicaltrials.gov/archive/NCT01217008 (Accessed 16 June 2011).
6. Lo B, Zettler P, Cedars MI, *et al.* (2005) A New Era in the Ethics of Human Embryonic Stem Cell Research. *Stem Cells* **23:** 1454–1459.
7. Illes J, Raimer JC, Kwon BK. (2011) Stem Cell Clinical Trials for Spinal Cord Injury: Readiness, Reluctance, Redefinition. *Stem Cell Review and Reports,* Online First, doi: 10.1007/S12015-011-9259-1.
8. Kumar AA, Kumar SR, Narayanan R, *et al.* (2009) Autologous Bone Marrow Derived Mononuclear Cell Therapy for Spinal Cord Injury: A Phase I/II Clinical Safety and Primary Efficacy Data. *Experimental and Clinical Transplantation* **7(4):** 241–48; Pal R, Venkataramana NK, Bansal A, *et al.* (2009) *Ex Vivo* Expanded Autologous Bone Marrow-derived Mesenchymal Stromal Cells in Human Spinal Cord Injury/Paraplegia: A Pilot Clinical Study. *Cytotherapy* **11(7):** 897–911; Cristante AF, Barros-Filho TE, Tatsui, N, *et al.* (2009) Stem Cells in the Treatment of Chronic Spinal Cord Injury: Evaluation of Somatosensitive Evoked Potentials in 39 Patients. *Spinal Cord* **47(10):** 733–738.
9. Lima C, Escada P, Pratas-Vital J, *et al.* (2010) Olfactory Mucosal Autografts and Rehabilitation for Chronic Traumatic Spinal Cord Injury. *Neurorehabilitation and Neural Repair* **24(1):** 10–22.
10. Mackay-Sim A, Féron F, Cochrane J, *et al.* (2008) Autologous Olfactory Ensheathing Cell Transplantation in Human Paraplegia: A 3-year Clinical Trial. *Brain* **131(9):** 2376–2386.
11. Macklin R. (2003) Bioethics, Vulnerability and Protection. *Bioethics* **17:** 472–486.

12. Kishi Y, Robinson RG, Forrester AW. (1995) Comparison Between Acute and Delayed Onset Major Depression After Spinal Cord Injury. *Journal of Nervous and Mental Disease* **183:** 286–292; Kennedy P, Rogers BA. (2000) Anxiety and Depression After Spinal Cord Injury: A Longitudinal Analysis. *Archives of Physical Medicine and Rehabilitation* **81:** 932–937; however, compare this with the results of Fuhrer MJ, Rintala DH, Hart KA, *et al.* (1993) Depressive Symptomatology in Persons with Spinal Cord Injury who Reside in the Community. *Archives of Physical Medicine and Rehabilitation* **74:** 255–260.
13. Bretzner F, Gilbert F, Baylis F, Brownstone RM. (2011) Target Populations for First-in-human Embryonic Stem Cell Research in Spinal Cord Injury. *Cell Stem Cell* **8:** 468–475.
14. Frank RG, Elliott TR, Corcoran JR, Wonderlich SA. (1987). Depression after Spinal Cord Injury: Is it Necessary? *Clinical Psychology Review* **7:** 611– 630; Elliott TR, Frank RG. (1996) Depression Following Spinal Cord Injury. *Archives of Physical Medicine and Rehabilitation* **77:** 816–823.
15. Kishi Y, Robinson RG, Forrester AW. (1995) Comparison Between Acute and Delayed Onset Major Depression After Spinal Cord Injury. *Journal of Nervous and Mental Disorder* **183:** 286–92; Dryden DM, Saunders LD, Rowe BH, *et al.* (2004) Utilization of Health Services Following Spinal Cord Injury: A 6-year Follow-up Study. *Spinal Cord* **42:** 513–525.
16. Robins L, Helzer J, Weissman M, *et al.* (1984) Lifetime Prevalence of Specific Psychiatric Disorders in Three Sites. *Archives of General Psychiatry* **41:** 949–958; Bland RC, Newman SC. (1988). Lifetime Prevalence of Psychiatric Disorders in Edmonton. *American Journal of Psychiatry* **77(suppl 338):** 24–32.
17. Frank RG, Elliott TR, Corcoran JR, Wonderlich SA. (1987) Depression after Spinal Cord Injury: Is it Necessary? *Clinical Psychology Review* **7:** 611– 630.
18. Dryden DM, Saunders LD, Rowe BH, *et al.* (2005) Depression Following Traumatic Spinal Cord Injury. *Neuroepidemiology* **25:** 55–61; Bombardier CH, Stroud MW, Esselman PC, Rimmele CT. (2004) Do Preinjury Alcohol Problems Predict Poorer Rehabilitation Progress in Persons with Spinal Cord Injury? *Archives of Physical Medicine and Rehabilitation* **85:** 1488–1492.
19. Dryden DM, Saunders LD, Rowe BH, *et al.* (2005) Depression Following Traumatic Spinal Cord Injury. *Neuroepidemiology* **25:** 55–61.
20. Dryden DM, Saunders LD, Rowe BH, *et al.* (2005) Depression Following Traumatic Spinal Cord Injury. *Neuroepidemiology* **25:** 55–61.
21. Burns AS, Lee BS, Ditunno JF, Jr., Tessler A. (2003) Patient Selection for Clinical Trials: The Reliability of the Early Spinal Cord Injury Examination. *Journal of Neurotrauma* **20:** 477–482.

22. Anderson JA, Kimmelman, J. (2010) Extending Clinical Equipoise to Phase 1 Trials Involving Patients: Unresolved Problems. *Kennedy Institute of Ethics Journal* **20:** 75–98.
23. Miller FG, Rosenstein DL. (2003) The Therapeutic Orientation to Clinical Trials. *New England Journal of Medicine* **348:** 1383–1386.
24. Anderson JA, Kimmelman J. (2010). Extending Clinical Equipoise to Phase 1 Trials Involving Patients: Unresolved Problems. *Kennedy Institute of Ethics Journal* **20:** 75–98.
25. Kimmelman J. (2007). The Therapeutic Misconception at 25: Treatment, Research, and Confusion. *Hastings Center Report* **37:** 36–42.
26. Kimmelman J, London AJ. (2011). Predicting Harms and Benefits in Translational Trials: Ethics, Evidence, and Uncertainty. *PLoS Medicine*, 8: e1001010.
27. Illes J, Raimer JC, Kwon BK. (2011) Stem Cell Clinical Trials for Spinal Cord Injury: Readiness, Reluctance, Redefinition. *Stem Cell Review and Reports* Online First, doi: 10.1007/S12015-011-9259-1.
28. Joffe S, Cook EF, Cleary PD, *et al.* (2001) Quality of Informed Consent in Cancer Clinical Trials: A Cross-sectional Survey. *Lancet* **358:** 1771–1777; Illes J, Raimer JC, Kwon BK. (2011) Stem Cell Clinical Trials for Spinal Cord Injury: Readiness, Reluctance, Redefinition. *Stem Cell Review and Reports*, Online First, doi: 10.1007/S12015-011-9259-1.
29. Feudtner C. (2008) Ethics in the Midst of Therapeutic Evolution. *Archives of Pediatric Adolescent Medicine* **162(9):** 854–857; Illes J, Raimer JC, Kwon BK. (2011) Stem Cell Clinical Trials for Spinal Cord Injury: Readiness, Reluctancc, Redefinition. *Stem Cell Review and Reports*, Online First, doi: 10.1007/S12015-011-9259-1.
30. Tetzlaff W, Okon E, Karimi-Abdolrezaee S, *et al.* (2010). A Systematic Review of Cellular Transplantation Therapies for Spinal Cord Injury. *Journal of Neurotrauma* **27:** e1001010.
31. Bretzner F, Gilbert F, Baylis F, Brownstone RM. (2011) Target Populations for First-in-human Embryonic Stem Cell Research in Spinal Cord Injury. *Cell Stem Cell* **8:** 468–475, pp. 469-470.
32. Illes J, Raimer JC, Kwon BK. (2011) Stem Cell Clinical Trials for Spinal Cord Injury: Readiness, Reluctance, Redefinition. *Stem Cell Review and Reports*, Online First, doi: 10.1007/S12015-011-9259-1.
33. Illes J, Raimer JC, Kwon BK. (2011) Stem Cell Clinical Trials for Spinal Cord Injury: Readiness, Reluctance, Redefinition. *Stem Cell Review and Reports*, Online First, doi: 10.1007/S12015-011-9259-1.

34. Tetzlaff W, Okon E, Karimi-Abdolrezaee S, *et al.* (2010). A Systematic Review of Cellular Transplantation Therapies for Spinal Cord Injury. *Journal of Neurotrauma* **27:** e1001010.
35. Buchanan AE, Brock DW. (1989) *Deciding for Others: The Ethics of Surrogate Decision Making.* Cambridge University Press, Cambridge.
36. Harris J. (2005). Scientific Research is a Moral Duty. *Journal of Medical Ethics* **31:** 242–248.
37. For this argument relating specifically to research on children see Harris J, Holm S. (2003) Should we Presume Moral Turpitude in our Children — Small Children and Consent to Medical Research *Theoretical Medicine* **24(2):** 121–129.
38. Hart HLA. (1955). Are There any Natural Rights? *The Philosophical Review* **64:** 175–191; Rawls J. (1972) *A Theory of Justice.* Harvard University Press, Cambridge, MA.
39. Illes J, Raimer JC, Kwon BK. (2011) Stem Cell Clinical Trials for Spinal Cord Injury: Readiness, Reluctance, Redefinition. *Stem Cell Review and Reports,* Online First, doi: 10.1007/S12015-011-9259-1.
40. Shakespeare W. *Julius Caesar* Act 4 Scene 2. Wells S, Taylor G (eds.), (1988) *The Oxford Shakespeare — Compact Edition.* Clarendon Press, Oxford.
41. Shakespeare W. *Julius Caesar* Act 4 Scene 2. Wells S, Taylor G (eds.), (1988) *The Oxford Shakespeare — Compact Edition.* Clarendon Press, Oxford.

Bibliography

Literature

A

Agar N. (2007) Embryonic Potential and Stem Cell. *Bioethics* **21:** 198–207.

Alghrani A. (2009) The Human Fertilisation and Embryology Act 2008: A Missed Opportunity? *Journal of Medical Ethics* **35(12):** 718–719.

Alvarez Manninen B. (2008) Are Human Embryos Kantian Persons?: Kantian Considerations in Favor of Embryonic Stem Cell Research. *Philosophy, Ethics, and Humanities in Medicine* **3(4)**.

Anderson JA, Kimmelman J. (2010) Extending Clinical Equipoise to Phase 1 Trials Involving Patients: Unresolved Problems. *Kennedy Institute of Ethics Journal* **20:** 75–98.

Anderssen N, Amlie C, Ytterøy EA. (2002) Outcomes for Children with Lesbian or Gay Parents: A Review of Studies from 1978 to 2000. *Scandinavian Journal of Psychology* **43(4):** 335–351.

Annas GJ. (1999) Waste and Longing — the Legal Status of Placental-Blood Banking. *Legal Issues in Medicine* **340(19):** 1521–1524.

Ashcroft RE. (2005) Making Sense of Dignity. *Journal of Medical Ethics* **31:** 679–682.

B

Barratt CLR, St John JC, Afnan M. (2004) Clinical Challenges in Providing Embryos for Stem-Cell Initiatives. *The Lancet* **364:** 115–118.

Beauchamp TL. (1999) The Failure of Theories of Personhood. *Kennedy Institute of Ethics Journal* **9(4):** 309–324.

Beyhan Z, Lager AE, Cibelli, JB. (2007) Interspecies Nuclear Transfer: Implications for Embryonic Stem Cell Biology. *Cell Stem Cell* **1(5):** 502–512.

Beyleveld D, Brownsword R. (2001) *Human Dignity in Bioethics and Biolaw.* Oxford University Press, Oxford — 2004, Reprint edition.

Blake D, Proctor M, Johnson N, Olive D. (2004) The Merits of Blastocyst Versus Cleavage Stage Embryo Transfer: A Cochrane Review. *Human Reproduction* **19(9):** 2174.

Bland RC, Newman SC. (1988) Lifetime Prevalence of Psychiatric Disorders in Edmonton. *American Journal of Psychiatry* **77(suppl 338):** 24–32.

Bombardier CH, Stroud MW, Esselman PC, Rimmele CT. (2004) Do Preinjury Alcohol Problems Predict Poorer Rehabilitation Progress in Persons with Spinal Cord Injury? *Archives of Physical Medicine and Rehabilitation* **85:** 1488–1492.

Bortolotti L, Harris J. (2006) Embryos and Eagles: Symbolic Value in Research and Reproduction. *Cambridge Quarterly of Health Care Ethics* **15:** 22–34.

Brazier M. (1998) Embryos' 'Rights': Abortion and Research. In: Freeman MDA (ed), *Medicine, Ethics and the Law.* Stevens & Sons, London.

Bretzner F, Gilbert F, Baylis F, Brownstone RM. (2011) Target Populations for First-in-human Embryonic Stem Cell Research in Spinal Cord Injury. *Cell Stem Cell* **8:** 468–75.

Brock DW. (2010) Creating Embryos for Use in Stem Cell Research. *Journal of Law & Medical Ethics* **38(2):** 229–37.

Brownsword R. (2004) Regulating Human Genetics: New Dilemmas for a New Millennium. *Medical Law Review* **12(1):** 14–39.

Buchanan AE, Brock DW. (1989) *Deciding for Others: The Ethics of Surrogate Decision Making.* Cambridge University Press, Cambridge.

Burns AS, Lee BS, Ditunno JF, Jr., Tessler A. (2003) Patient Selection for Clinical Trials: The Reliability of the Early Spinal Cord Injury Examination. *Journal of Neurotrauma* **20:** 477–82.

Burt RK, Loh Y, Cohen B, *et al.* (2009) Autologous Non-Myeloablative Haemopoietic Stem Cell Transplantation in Relapsing-Remitting Multiple Sclerosis: A Phase I/II Study. *The Lancet Neurology* **8:** 244–253.

Byrne JA, Pederson DA, Clepper LL, *et al.* (2007) Producing Primate Embryonic Stem Cells by Somatic Cell Nuclear Transfer. *Nature* **450(7169):** 497–502.

C

Capron AM. (2001) Stem Cells: Ethics, Law and Politics. *Biotechnology Law Report* **20:** 678–699.

Carroll L. (1865/1998) *Alice's Adventures in Wonderland.* Penguin Classics, London.

Casper MJ. (1998) *The Making of the Unborn Patient: A Social Anatomy of Fetal Surgery.* Rutgers University Press, London.

Chan S, Harris J. (2008) Adam's Fibroblast? The (pluri)potential of iPCs. *Journal of Medical Ethics* **34(2):** 65–66.

Chen HF, Kuo HC, Chien CL, *et al.* (2007) Derivation, Characterization, and Differentiation of Human Embryonic Stem Cells: Comparing Serum-containing Versus Serum-free Media and Evidence of Germ Cell Differentiation. *Human Reproduction* **22(2):** 567–577.

Choudary M, Haimes E, Herbert M, *et al.* (2004) Demographic, Medical and Treatment Characteristics Associated With Couples' Decisions to Donate Fresh Spare Embryos for Research. *Human Reproduction* **19(9):** 2091–2096.

Clark AT, Bodnar MS, Fox M, *et al.* (2004) Spontaneous Differentiation of Germ Cells from Human Embryonic Stem Cells *In Vitro. Human Molecular Genetics* **13(7):** 727–739.

Conley JJ. (2002) Delayed Animation: An Ambiguity and Its Abuses. In: Koterski JW (ed), *Life and Learning XII: Proceedings of the Twelfth University Faculty for Life Conference at Ave Maria Law School.* Available at http://uffl.org/vol12/conley12.pdf

Cristante AF, Barros-Filho TE, Tatsui N, *et al.* (2009) Stem Cells in the Treatment of Chronic Spinal Cord Injury: Evaluation of Somatosensitive Evoked Potentials in 39 Patients. *Spinal Cord* **47(10):** 733–8.

D

De Beaufort I, English V. (2000) Between Pragmatism and Principles: On the Morality of Using the Results of Research that a Country Considers Immoral. In: Gunning J (ed), *Assisted Conception: Research, Ethics, and Law.* Ashgate, Dartmouth, pp. 57–65.

Devaney S. (2010) Tissue Providers for Stem Cell Research: The Dispossessed. *Law, Innovation and Technology* **2(2):** 165–191.

Devolder K, Harris J. (2005) Compromise and Moral Complicity in the Embryonic Stem Cell Debate. In: Athanassoulis N (ed), *Philosophical Reflections on Medical Ethics.* Palgrave MacMillan, Hampshire, pp. 88–108.

Devolder K. (2005) Advance Directives to Protect Embryos. *Journal of Medical Ethics* **31:** 497–8.

Devolder K. (2006) What's in a Name? Embryos, Entities, and ANTities in the Stem Cell Debate. *Journal of Medical Ethics* **32:** 43–48.

Devolder K, Harris J. (2007) The Ambiguity of the Embryo: Ethical Inconsistency in the Human Embryonic Stem Cell Debate. *Metaphilosophy* **38**: 153–169.

Devolder K. (2009) To Be, or Not to Be? *EMBO Reports* **10(12):** 1285–1877.

Devolder K. (2010) Complicity in Stem Cell Research: The Case of Induced Pluripotent Stem Cells. *Human Reproduction* **25(9):** 2175–2180.

Dolgin E. (2010) Gene Flaw Found in Induced Stem Cells. *Nature* **464:** 663.

Douglas T. (2008) Moral Enhancement. *Journal of Applied Philosophy* **25(3):** 228–245.

Drusenheimer N, Wulf G, Nolte J, *et al.* (2007) Putative Human Male Germ Cells from Bone Marrow Stem Cells. *Society of Reproduction & Fertility* **63(suppl):** 69–76.

Dryden DM, Saunders LD, Rowe BH, *et al.* (2004) Utilization of Health Services Following Spinal Cord Injury: A 6-year Follow-up Study. *Spinal Cord* **42:** 513–25.

Dryden DM, Saunders LD, Rowe BH, *et al.* (2005) Depression Following Traumatic Spinal Cord Injury. *Neuroepidemiology* **25:** 55–61.

Dyce PW, Wen L, Li J. (2006) *In Vitro* Germline Potential of Stem Cells Derived from Fetal Porcine Skin. *Nature Cell Biology* **8(4):** 384–390.

E

Easton L. (2003) Man and Superman. *British Medical Journal* **326:** 1287–1290.

Eberl JT. (2000) The Beginning of Personhood: A Thomistic Biological Analysis. *Bioethics* **14(2):** 134–157.

Ehrich K, Williams C, Farsides B. (2008) The Embryo as Moral Work Object: PGD/IVF Staff Views and Experiences. *Sociology of Health and Illness* **30:** 772–787.

Ehrich K, Williams C, Farsides B. (2010) Fresh or Frozen? Classifying 'Spare' Embryos for Donation to Human Embryonic Stem Cell Research. *Social Science and Medicine* **71:** 2204–2211.

Elliott TR, Frank RG. (1996) Depression Following Spinal Cord Injury. *Archives of Physical Medicine and Rehabilitation* **77:** 816–23.

English J. (1975) Abortion and the Concept of a Person. *Canadian Journal of Philosophy* **5:** 236.

Mertes H, Pennings G. (2010) Ethical Aspects of the Use of Stem Cell Derived Gametes for Reproduction. *Health Care Analysis* **18(3):** 267–278.

Ethics Committee of the American Society for Reproductive Medicine. (2002) Donating Spare Embryos for Embryonic Stem Cell Research. *Fertility and Sterility* **78(5):** 957–960.

F

Farsides B, Williams C, Alderson P. (2004) Aiming Towards 'Moral Equilibrium': Health Care Professionals' Views of Working within the Morally Contested Field of Antenatal Screening. *Journal of Medical Ethics* **30:** 505–509.

Feinberg J. (1985) Comment: Wrongful Conception and the Right Not To Be Harmed. *Harvard Journal of Law and Public Policy* **8:** 57–77.

Feudtner C. (2008) Ethics in the Midst of Therapeutic Evolution. *Archives of Pediatric Adolescent Medicine* **162(9):** 854–7.

Findlay JK, Gear ML, Illingworth PJ, *et al.* (2007) Human Embryo: A Biological Definition. *Human Reproduction* **22:** 905–911.

Finnis J. (1997) The Good of Marriage and the Morality of Sexual Relations: Some Philosophical and Historical Observations. *American Journal of Jurisprudence* **42:** 97–134.

Fletcher JC. (2001) NBAC's Arguments on Embryo Research: Strengths and Weaknesses. In: Holland S, Lebacqz K, Zoloth L (eds), *The Human Embryonic Stem Cell Debate: Science, Ethics, and Public Policy.* MIT Press, Cambridge, MA, pp. 61–72.

Ford M. (2008) Nothing and Not-nothing: Law's Ambivalent Response to Transformation and Transgression at the Beginning of Life. In: Smith SW, Deazley R (eds), *The Legal, Medical and Cultural Regulation of the Body.* Ashgate, Surrey, pp. 21–46.

Fox M. (2000) Pre-persons, Commodities or Cyborgs: The Legal Construction and Representation of the Embryo. *Health Care Analysis* **8:** 171–188.

Fox M. (2008) Legislating Interspecies Embryos. In: Smith SW, Deazley R (eds), *The Legal, Medical and Cultural Regulation of the Body.* Ashgate, Surrey, pp. 95–126.

Frank RG, Elliott TR, Corcoran JR, Wonderlich SA. (1987) Depression After Spinal Cord Injury: Is it Necessary? *Clinical Psychology Review* **7:** 611– 630.

Franklin S. (2006) *Born and Made: An Ethnography of Preimplantation Genetic Diagnosis.* Princeton University Press, Princeton.

Franklin S. (2006) Embryo Economies: The Double Reproductive Value of Stem Cells. *Biosocieties* **1:** 71–90.

Franklin S. (2010) Response to Marie Fox and Thérèse Murphy. *Social & Legal Studies* **19(4):** 505–510.

Friedman A. (2009) Intransitive Ethics. *Journal of Moral Philosophy* **6:** 277–297.

Fuhrer MJ, Rintala DH, Hart KA, *et al.* (1993) Depressive Symptomatology in Persons with Spinal Cord Injury Who Reside in the Community. *Archives of Physical Medicine and Rehabilitation* **74:** 255–60.

Fukuyama F. (2002) *Our Posthuman Future: Consequences of the Biotechnology Revolution.* Profile Books, New York.

G

Gardner J. (2004) Book review of Kutz, C., *Complicity, Ethics, and Law for a Collective Age.* Cambridge University Press, Cambridge. *Ethics* 827–830.

Gardner J. (2007) Complicity and Causality. *Criminal Law and Philosophy* **1:** 127–41.

Geijsen N, Horoschak M, Kim K, *et al.* (2004) Derivation of Embryonic Germ Cells and Male Gametes from Embryonic Stem Cells. *Nature* **427(6970):** 148–154.

Geoghegan E, Byrnes L. (2008) Mouse Induced Pluripotent Stem Cells. *International Journal of Developmental Biology* **52(8):** 1015–1022.

George RP, Lee P. (Fall 2004/Winter 2005) Embryos and Acorns. *The New Atlantis* **7:** 90–100.

George RP, Tollefsen C. (2008) *Embryo: A Defense of Human Life.* The Doubleday Broadway Publishing Group, New York.

Geron (2010) Phase 1 Safety Study of GRNOPC1 in Patients with Neurologically Complete, Subacute, Spinal Cord Injury. Available at http://clinicaltrials.gov/archive/NCT01217008 (Accessed 16 June 2011).

Gillam L. (1997) Arguing by Analogy in the Fetal Tissue Debate. *Bioethics* **11(5):** 397–412.

Green RM (ed). (2000) *Encyclopedia of Ethical, Legal and Policy Issues in Biotechnology* (Vol. 2). Wiley-Interscience, New York.

Green R. (2002) Benefiting from 'Evil': An Incipient Moral Problem in Human Stem Cell Research. *Bioethics* **16:** 544–556.

Grisez G. (1993) *The Way of the Lord Jesus, Vol. 2: Living a Christian Life.* Franciscan Press, Quincy, IL.

Grubb A. (1998) 'I, Me Mine': Bodies, Parts and Property. *Medical Law International* **3:** 299–317.

Guenin LM. (2004) On Classifying the Developing Organism. *Connecticut Law Review* **36:** 1115–1131.

H

Haimes E, Porz R, Scully J, Rehmann-Sutter C. (2008) "So What is an Embryo?" A Comparative Study of the Views of Those Asked to Donate Embryos for hESC Research in the UK and Switzerland. *New Genetics and Society* **27(2):** 113–126.

Halsbury's *Laws of England — Personal Property*, Volume 35 (5th edition). LexisNexis, London 2011, p. 1225.

Hammarberg K, Tinney L. (2006) Deciding the Fate of Supernumerary Frozen Embryos: A Survey of Couples' Decisions and the Factors Influencing Their Choice. *Fertility and Sterility* **86(1):** 86–91.

Hammersley M. (2002) Ethnography and Realism. In: Huberman MA, Miles MB (eds), *The Qualatative Researchers Companion.* Sage Publications, California & London, pp. 65–80.

Hanna J, *et al.* (2007) Treatment of Sickle Cell Anemia Mouse Model with iPS Cells Generated from Autologous Skin. *Science* **318:** 1920–23.

Harris J. (1985) *The Value of Life: An Introduction to Medical Ethics.* Routledge & Kegan Paul, London.

Harris J. (1999) The Concept of the Person and the Value of Life. *Kennedy Institute of Ethics Journal* **9(4):** 293–308.

Harris J. (2003) Stem Cells, Sex, and Procreation. *Cambridge Quarterly of Health Care Ethics* **12:** 353–371.

Harris J, Holm, S. (2003) Should We Presume Moral Turpitude in Our Children — Small Children and Consent to Medical Research. *Theoretical Medicine* **24(2):** 121–29.

Harris J. (2004) *On Cloning.* Routledge, London.

Harris J. (2005) Scientific Research is a Moral Duty. *Journal of Medical Ethics* **31:** 242–48.

Harris J. (2007) *Enhancing Evolution.* Princeton University Press, Princeton and Oxford.

Harris J. (2008) Global Norms, Informed Consensus, and Hypocrisy in Bioethics. In: Green RM, Donovan A, Jauss SA (eds), *Global Bioethics — Issues of Conscience for the Twenty-first Century.* Oxford University Press, Oxford.

Harris J. (2011) Taking the 'Human' Out of Human Rights. *Cambridge Quarterly of Healthcare Ethics* **20(1):** 9–20.

Hart HLA. (1955) Are There Any Natural Rights? *The Philosophical Review* **64:** 175–91.

Heinemann T, Honnefelder L. (2002) Principles of Ethical Decision-making Regarding Embryonic Stem Cell Research in Germany. *Bioethics* **16:** 530–43.

Heng BC. (2006) Donation of Surplus Frozen Embryos for Stem Cell Research or Fertility Treatment — Should Medical Professionals and Healthcare Institutions Be Allowed to Exercise Undue Influence on the Informed Decision of Their Former Patients? *Journal of Assisted Reproduction and Genetics* **23:** 381–382.

Hennette-Vauchez S. (2009) Words Count: How Interest in Stem Cells has Made the Embryo Available: A Look at the French Law on Bioethics. *Medical Law Review* **17:** 52–75.

Hochedlinger K, Jaenisch R. (2006) Nuclear Reprogramming and Pluripotency. *Nature* **441(7097):** 1061–1067.

Holland S. (2001) Beyond the Embryo: A Feminist Appraisal of the Embryonic Stem Cell Debate. In: Holland S, Lebacqz K, Zoloth L (eds), *The Human Embryonic Stem Cell Debate: Science, Ethics and Public Policy.* MIT Press, Cambridge, MA, pp. 73–86.

Holland S. (2001) Contested Commodities at Both Ends of Life: Buying and Selling Gametes, Embryos, and Body Tissues. *Kennedy Institute of Ethics Journal* **11(3):** 263–284.

Holm S. (2003) The Ethical Case Against Stem Cell Research. *Cambridge Quarterly of Health Care Ethics* **12(4):** 372–83.

Holm S. (2006) Are Countries that Ban Human Embryonic Stem Cell Research Hypocritical? *Regenerative Medicine* **1:** 357–359.

Holm S. (2008) Time to Reconsider Stem Cell Ethics: The Importance of Induced Pluripotent Cells. *Journal of Medical Ethics* **34:** 63–64.

Honoré AM. (1959, 1985) *Causation in the Law* (2nd Edition). Clarendon Press, Oxford.

Honoré AM. (1987) *Making Law Bind: Essays Legal and Philosophical.* Clarendon Press, Oxford.

Honoré AM. (2010) Causation in the Law. In: Zalta EN (ed), *The Stanford Encyclopedia of Philosophy (Winter 2010 Edition).* Available at http://plato.stanford.edu/archives/win2010/entries/causation-law/

Hubner K, Fuhrmann G, Christenson LK, *et al.* (2003) Derivation of Oocytes from Mouse Embryonic Stem Cells. *Science* **300(5623):** 1251–1256.

Hume D. (1978) *A Treatise of Human Nature* (2nd Edition). Clarendon Press, Oxford.

Hwang WS, Ryu YJ, Park JH, *et al.* (2004) Evidence of a Pluripotent Human Embryonic Stem Cell Line Derived from a Cloned Blastocyst. *Science* **303(5664):** 1669–1674.

Hwang WS, *et al.* (2005) Patient-specific Embryonic Stem Cells Derived from Human SCNT Blastocysts. *Science* **308(5729):** 1777–1783.

I

Illes J, Raimer JC, Kwon BK. (2011) Stem Cell Clinical Trials for Spinal Cord Injury: Readiness, Reluctance, Redefinition. *Stem Cell Review and Reports*, Online First, doi: 10.1007/S12015-011-9259-1.

Isasi RM, Knoppers BM. (2007) Monetary Payments for the Procurement of Oocytes for Stem Cell Research: In Search of Ethical and Political Consistency. *Stem Cell Research* **1:** 37–44.

J

Jackson E. (2006) Fraudulent Stem Cell Research and Respect for the Embryo. *Biosocieties* **1:** 349–356.

Jacob M, Prainsack B. (2010) Embryonic Hopes: Controversy, Alliance, and Reproductive Entities in Law and the Social Sciences. *Social & Legal Studies* **19:** 497–517.

Joffe S, Cook EF, Cleary PD, *et al.* (2001) Quality of Informed Consent in Cancer Clinical Trials: A Cross-sectional Survey. *Lancet* **358:** 1771–1777.

Johnson MH. (2006) Escaping the Tyranny of the Embryo? A New Approach to ART Regulation Based on UK and Australian Experiences. *Human Reproduction* **21(11):** 2756–2765.

Jonsen A. (1998) *The Birth of Bioethics.* Oxford University Press, New York.

Judicial Studies Board. (2008) *Guidelines for the Assessment of General Damages in Personal Injury Cases* (9th Edition). Oxford University Press, Oxford.

K

Kaczor C. (2005) *The Edge of Life: Human Dignity and Contemporary Bioethics.* Springer, Dordrecht.

Kadish S. (1987) A Theory of Complicity. In: Gavison R (ed), *Issues in Contemporary Legal Philosophy: The Influence of H.L.A. Hart.* Clarendon Press, Oxford, pp. 287–303.

Kain P. (2009) Kant's Defense of Human Moral Status. *Journal of the History of Philosophy* **47(1):** 59–102.

Kant I. (1785) Groundwork of the Metaphysics of Morals. White Beck L (trans). *Foundations of the Metaphysics of Morals.* Library of Liberal Art: 1959 (VI).

Kant I. (1785/1997) Groundwork of the Metaphysics of Morals. Gregor MJ (trans). Cambridge University Press, Cambridge.

Karpin I. (2006) The Uncanny Embryos: Legal Limits to the Human and Reproduction Without Women. *Sydney Law Review* **28:** 599–623.

Kass L. (1997) The Wisdom of Repugnance. *The New Republic* **216(22)**.

Kennedy D. (2006) Editorial Retraction. *Science* **311(5759):** 335.

Kennedy P, Rogers BA. (2000) Anxiety and Depression After Spinal Cord Injury: A Longitudinal Analysis. *Archives of Physical Medicine and Rehabilitation* **81:** 932–37.

Kerkis AA, Fonseca SASAS, Serafim RCRC, *et al.* (2007) *In Vitro* Differentiation of Male Mouse Embryonic Stem Cells into Both Presumptive Sperm Cells and Oocytes. *Cloning and Stem Cells* **9(4):** 535–48.

Kiesling A. (2004) What is an Embryo? *Connecticut Law Review* **36:** 1051–1092.

Kimmelman J. (2007) The Therapeutic Misconception at 25: Treatment, Research, and Confusion. *Hastings Center Report* **37:** 36–42.

Kimmelman J, London AJ. (2011) Predicting Harms and Benefits in Translational Trials: Ethics, Evidence, and Uncertainty. *PLoS Medicine* **8:** e1001010.

King P. (1978–79) The Juridical Status of the Fetus: A Proposal for Legal Protection of the Unborn. *Michigan Law Review* **77:** 1647–1687.

Kishi Y, Robinson RG, Forrester AW. (1995) Comparison between Acute and Delayed Onset Major Depression after Spinal Cord Injury. *Journal of Nervous and Mental Disorder* **183:** 286–92.

Knowles LP. (1999) Human Primordial Stem Cells: Property, Progeny and Patents. *Hastings Center Report* **29(2):** 38.

Korobkin R. (2007) Buying and Selling Human Tissues for Stem Cell Research. *Arizona Law Review* **49:** 45–67.

Kumar AA, Kumar SR, Narayanan R, *et al.* (2009) Autologous Bone Marrow Derived Mononuclear Cell Therapy for Spinal Cord Injury: A Phase I/II Clinical Safety and Primary Efficacy Data. *Experimental and Clinical Transplantation* **7(4):** 241–48.

Kutz C. (2000) *Complicity, Ethics, and Law for a Collective Age.* Cambridge University Press, Cambridge.

Kutz C. (2007) Causeless Complicity. *Criminal Law and Philosophy* **1:** 289–305.

L

Lacham-Kaplan O, Chy H, Trounson A. (2006) Testicular Cell Conditioned Medium Supports Differentiation of Embryonic Stem Cells into Ovarian Structures Containing Oocytes. *Stem Cells* **24(2):** 266–273.

Lako M, Armstrong L, Stojkovic M. (2010) Induced Pluripotent Stem Cells: It Looks Simple but Can Looks Deceive? *Stem Cells Express* **28(5):** 845–850.

Laurance J. (2010) British Boy Receives Trachea Transplant Built with His Own Stem Cells. *British Medical Journal* **340:** 1633.

Lebacqz K. (2001) Who 'Owns' Cells and Tissues? *Health Care Analysis* **9:** 353–367.

Lee P. (2008) Marriage, Procreation, and Same-sex Unions. *The Monist* **91(3–4):** 422–438.

Lerou PH, Yabuuchi A, Huo H, *et al.* (2008) Human Embryonic Stem Cell Derivation from Poor-quality Embryos. *Nature Biotechnology* **26:** 212–214.

Lima C, Escada P, Pratas-Vital J, *et al.* (2010) Olfactory Mucosal Autografts and Rehabilitation for Chronic Traumatic Spinal Cord Injury. *Neurorehabilitation and Neural Repair* **24(1):** 10–22.

Lindsay RA. (2005) Slaves, Embryos, and Nonhuman Animals: Moral Status and the Limitations of Common Morality Theory. *Kennedy Institute of Ethics Journal* **15:** 323–346.

Little MO. (1998) Cosmetic Surgery, Suspect Norms, and the Ethics of Complicity. In: Parens E (ed), *Enhancing Human Traits: Ethical and Social Implications.* Georgetown University Press, Washington, DC, pp. 162–176.

Little MO. (2008) Abortion and the Margins of Personhood. *Rutgers Law Journal* **39:** 331–348.

Lo B, Zettler P, Cedars MI, *et al.* (2005) A New Era in the Ethics of Human Embryonic Stem Cell Research. *Stem Cells* **23:** 1454–59.

Locke J. (1690/1999) *An Essay Concerning Human Understanding.* Ch.XXVII, Bk. II: 217. Available at http://www2.hn.psu.edu/faculty/jmanis/locke/humanund.pdf

Lockwood M. (1985) *Moral Dilemmas in Modern Medicine.* Oxford University Press, Oxford.

Lockwood M. (2001) The Moral Status of the Human Embryo. *Human Fertility* **4(4):** 267–269.

Lovell-Badge R. (2008) The Regulation of Human Embryo and Stem-Cell Research in the United Kingdom. *Nature Reviews Molecular Cell Biology* **9:** 998–1003.

M

Mackay-Sim A, Féron F, Cochrane J, *et al.* (2008) Autologous Olfactory Ensheathing Cell Transplantation in Human Paraplegia: A 3-year Clinical Trial. *Brain* **131(9):** 2376–86.

Macklin R. (2003) Bioethics, Vulnerability and Protection. *Bioethics* **17:** 472–86.

Macklin R. (2003) Dignity is a Useless Concept. *British Medical Journal* **327:** 1419–1420.

Marquis D. (1989) Why Abortion is Immoral. *Journal of Philosophy* **86(4):** 183–202.

Marquis D. (2005) Savulescu's Objections to the Future of Value Argument. *Journal of Medical Ethics* **31:** 119–122.

Mathews D, Donovan PJ, Harris J, *et al.* (2009) Pluripotent Stem Cell-derived Gametes: Truth and Potential Consequences. *Cell Stem Cell* **5(1):** 11–14.

McHugh PR. (2004) Zygote and 'Clonote': The Ethical Use of Embryonic Stem Cells. *New England Journal of Medicine* **351:** 209–11.

McLeod C, Baylis F. (2007) Donating Fresh Versus Frozen Embryos to Stem Cell Research: In Whose Interests? *Bioethics* **21(9):** 465–477.

Menzies P. (2009) Counterfactual Theories of Causation. In: Zalta EN (ed), *The Stanford Encyclopedia of Philosophy (Fall 2009 Edition).* Available at http://plato.stanford.edu/archives/fall2009/entries/causation-counterfactual/

Mertes H, Pennings G. (2009) Stem Cell Research Policies: Who's Afraid of Complicity? *Reproductive Biomedicine Online* **19(suppl 1):** 38–42.

Mertes H, Pennings G. (2010) Ethical Aspects of the Use of Stem Cell Derived Gametes for Reproduction. *Health Care Analysis* **18(3):** 267–278.

Miller FG, Rosenstein DL. (2003) The Therapeutic Orientation to Clinical Trials. *New England Journal of Medicine* **348:** 1383–1386.

Montgomery J. (1991) Rights, Restraints and Pragmatism: The Human Fertilisation and Embryology Act 1990. *Modern Law Review* **54:** 524–534.

Mulkay M. (1994) The Triumph of the Pre-Embryo: Interpretations of the Human Embryo in Parliamentary Debate Over Embryo Research. *Social Studies of Science* **24:** 611–639.

Mulkay M. (1997) *The Embryo Research Debate.* Cambridge University Press, Cambridge.

Murray D. (2007) *What is a Person?* Veritas, Dublin.

N

Nakagawa M, Koyanagi M, Tanabe K, *et al.* (2008) Generation of Induced Pluripotent Stem Cells Without Myc from Mouse and Human Fibroblasts. *Nature Biotechnology* **26(1):** 101–106.

Nayernia K, Nolte J, Michelmann HW, *et al.* (2006) *In Vitro*-Differentiated Embryonic Stem Cells Give Rise to Male Gametes that Can Generate Offspring Mice. *Developmental Cell* **11(1):** 125–132.

Nussbaum M. (1995) Objectification. *Philosophy & Public Affairs* **24:** 249–291.

O

Ohinata Y, Ohta H, Shigeta M, *et al.* (2009) A Signaling Principle for the Specification of the Germ Cell Lineage in Mice. *Cell* **137(3):** 571–584.

Outka G. (2009) The Ethics of Embryonic Stem Cell Research and the Principle of Nothing is Lost. *Yale Journal of Health Policy & Ethics* **9:** 585–602.

P

Pal R, Venkataramana NK, Bansal A, *et al.* (2009) *Ex Vivo* Expanded Autologous Bone Marrow-derived Mesenchymal Stromal Cells in Human Spinal Cord Injury/Paraplegia: A Pilot Clinical Study. *Cytotherapy* **11(7):** 897–911.

Pardo R, Calvo F. (2008) Attitudes Towards Embryo Research, Worldviews and the Moral Status of the Embryo Frame. *Science Communications* **30:** 8–47.

Park I, Zhao R, West JA, *et al.* (2008) Reprogramming of Human Somatic Cells to Pluripotency with Defined Factors. *Nature* **451(7175):** 141–146.

Parry S. (2003) The Politics of Cloning: Mapping the Rhetorical Convergence of Embryos and Stem Cells in Parliamentary Debates. *New Genetics and Society* **22:** 177–200.

Pennings G, Van Steirteghem A. (2004) The Subsidiarity Principle in the Context of Embryonic Stem Cell Research. *Human Reproduction* **19(5):** 1060–1064.

Pennings G. (2000) What are the Ownership Rights for Gametes and Embryos? *Human Reproduction* **15(5):** 979–986.

Potter J. (2003) *Representing Reality: Discourse, Rhetoric and Social Construction.* Sage, London.

Prieur MR, Atkinson J, Hardingham L, *et al.* (2006) Stem Cell Research in a Catholic Institution: Yes or No? *Kennedy Institute of Ethics Journal* **16(1):** 73–98.

Provoost V, *et al.* (2009) Infertility Patients' Beliefs About Their Embryos and Their Disposition Preferences. *Human Reproduction* **1(1):** 1–10.

R

Rachels S. (2005) Counterexamples to the Transitivity of 'Better Than'. *Recent Work on Intrinsic Value — Library of Ethics and Applied Philosophy* **17(Part IV):** 249–263.

Ramsey P. (1961) *War and the Christian Conscience: How Shall Modern War Be Conducted Justly?* Duke University Press, Durham, NC.

Rawls J. (1972) *A Theory of Justice.* Harvard University Press, Cambridge, MA.

Resnik DB. (2002) The Commercialization of Human Stem Cells: Ethical and Policy Issues. *Health Care Analysis* **10(2):** 147.

Robertson JA. (1995) Symbolic Issues in Embryo Research. *Hastings Center Report* **25(1):** 37–38.

Robertson JA. (1999) Ethics and Policy in Embryonic Stem Cell Research. *Kennedy Institute of Ethics Journal* **9(2):** 109–136.

Robertson JA. (2004) Causative vs. Beneficial Complicity in the Embryonic Stem Cell Debate. *Connecticut Law Review* **36:** 1099–1113.

Robertson JA. (2010) Embryo Stem Cell Research: Ten Years of Controversy. *The Journal of Law, Medicine and Ethics* **38(2):** 191–203.

Robins L, Helzer J, Weissman M, *et al.* (1984) Lifetime Prevalence of Specific Psychiatric Disorders in Three Sites. *Archives of General Psychiatry* **41:** 949–58.

S

Salter B. (2007) Bioethics, Politics and the Moral Economy of Human Embryonic Stem Cell Science: The Case of the European Union's Sixth Framework Programme. *New Genetics and Society* **26:** 269–288.

Sandel M. (2007) *The Case Against Perfection: Ethics in the Age of Genetic Engineering.* Harvard University Press, Cambridge, MA.

Sandel M. (2004) Embryo Ethics: The Moral Logic of Stem-Cell Research. *New England Journal of Medicine* **351:** 207–209.

Savulescu J. (2002) Abortion, Embryo Destruction, and the Future of Value Argument. *Journal of Medical Ethics* **28:** 133–35.

Savulescu J. (2004) Embryo Research: Are There Any Lessons from Natural Reproduction? *Cambridge Quarterly of Health Care Ethics* **13:** 68–95.

Savulescu J. (2006) Genetic Interventions and the Ethics of Enhancement of Human Beings. In: Steinbock B (ed), *The Oxford Handbook of Bioethics.* Oxford University Press, Oxford.

Savulescu J. (2007) Future People, Involuntary Medical Treatment in Pregnancy and the Duty of Easy Rescue. *Utilitas* **19(1):** 1–20.

Schroeder D. (2008) Dignity: Two Riddles and Four Concepts. *Cambridge Quarterly of Health Care Ethics* **17:** 230–238.

Scott R, Williams C, Ehrich K, Farsides B. (2007) The Appropriate Extent of Pre-Implantation Genetic Diagnosis: Health Professionals' and Scientists Views on Requirements for 'Significant Risk of Serious Genetic Condition'. *Medical Law Review* **15:** 320–356.

Scott R. (2007) *Choosing Between Possible Lives: Law and Ethics of Prenatal and Preimplantation Genetic Diagnosis.* Hart Publishing, Oxford & Portland, OR.

Secrest AM. (2010) All-Cause Mortality Trends in a Large Population-Based Cohort With Long-Standing Childhood-Onset Type 1 Diabetes. *Diabetes Care* **33:** 2573–2579.

Shakespeare W. (1988) *Julius Caesar* (Act 4 Scene 2). In: Wells S, Taylor G (eds), *The Oxford Shakespeare — Compact Edition.* Clarendon Press, Oxford.

Shapiro HT. (2004) What is an Embryo? Commentary. *Connecticut Law Review* **36:** 1093–1097.

Skene L, Testa G, Hyun I, *et al.* (2009) Ethics Report on Interspecies Somatic Cell Nuclear Transfer Research. *Cell Stem Cell* **5(1):** 27–30.

Silver L. (1999) *Remaking Eden: Cloning, Genetic Engineering and the Future of Humankind?* Phoenix Giant, London.

Singer P. (1993) *Practical Ethics.* Cambridge University Press, Cambridge.

Sipp D. (2010) Challenges in the Clinical Application of Induced Pluripotent Stem Cells. *Stem Cell Research & Therapy* **1:** 9.

Smith SW. (2008) Precautionary Reasoning in Determining Moral Worth. In: Freeman MDA (ed), *Law and Bioethics.* Oxford University Press, Oxford.

Sommerville M. (2007) Children's Human Rights and Unlinking Child-parent Biological Bonds with Adoption, Same-sex marriage and New Reproductive Technologies. *Journal of Family Studies* **13(2):** 179–201.

Sperling S. (2004) Managing Potential Selves: Stem Cells, Immigrants, and German Identity. *Science and Public Policy* **31:** 139–149.

Sparman ML, Tachibana M, Mitalipov SM. (2010) Cloning of Non-human Primates: The Road 'Less Travelled By'. *International Journal of Developmental Biology* **54(11–12):** 1671–1678.

Stacey G, Hunt CJ. (2006) The UK Stem Cell Bank: A UK Government-Funded, International Resource Center for Stem Cell Research. *Regenerative Medicine* **1:** 139–142.

Strathern M. (1992) *Reproducing the Future: Anthropology, Kinship, and the New Reproductive Technologies.* Manchester University Press, Manchester.

Sumner LW. (1981) *Abortion and Moral Theory.* Princeton University Press, Princeton, NJ.

Sunstein CR. (1995) Incompletely Theorized Agreements. *Harvard Law Review* **108:** 1733–1772.

Szawarski Z. (1996) Talking About Embryos. In: Evans D (ed), *Conceiving the Embryo: Ethics, Law, and Practice in Human Embryology.* Nijhoff, The Hague, pp. 119–134.

T

Tada M, Takahama Y, Abe K, *et al.* (2001) Nuclear Reprogramming of Somatic Cells by *In Vitro* Hybridization with ES Cells. *Current Biology* **11(19):** 1553–1558.

Takahashi K, Tanabe K, Ohnuki M, *et al.* (2007) Induction of Pluripotent Stem Cells from Adult Human Fibroblasts by Defined Factors. *Cell* **131(5):** 861–872.

Takahashi K, Yamanaka S. (2006) Induction of Pluripotent Stem Cells from Mouse Embryonic and Adult Fibroblast Cultures by Defined Factors. *Cell* **126(4):** 663–676.

Takala T, Häyry M. (2007) Benefiting from Past Wrongdoing, Human Embryonic Stem Cell Lines, and the Fragility of the German Legal Position. *Bioethics* **21(3):** 150–159.

Taylor PL. (2010) Overseeing Innovative Therapy Without Mistaking it for Research: A Function Based Model Based on Old Truths, New Capacities and Lessons Learned from Stem Cells. *The Journal of Law, Medicine and Ethics* **38(2):** 286–302.

Temkin L. (1987) Intransitivity and the Mere Addition Paradox. *Philosophy and Public Affairs* **16:** 138–187.

Tesar PJ, Chenoweth JG, Brook FA, *et al.* (2007) New Cell Lines from Mouse Epiblast Share Defining Features with Human Embryonic Stem Cells. *Nature* **448(7150):** 196–199.

Testa G, Harris J. (2005) Ethics and Synthetic Gametes. *Bioethics* **19(2):** 146–166.

Tetzlaff W, Okon E, Karimi-Abdolrezaee S, *et al.* (2010) A Systematic Review of Cellular Transplantation Therapies for Spinal Cord Injury. *Journal of Neurotrauma* **27**, doi: 10.1089/neu.2009.1177.

Thompson C. (2005) *Making Parents: The Ontological Choreography of Reproductive Technologies.* The MIT Press, Massachusetts.

Thomson JA, Itskovitz-Eldor J, Shapiro SS, *et al.* (1998) Embryonic Stem Cell Lines Derived from Human Blastocysts. *Science* **282(5391):** 1145–1147.

Tooley M. (1983) *Abortion and Infanticide.* Clarendon Press, Oxford.

Toyooka Y, Tsunekawa N, Akasu R, *et al.* (2003) Embryonic Stem Cells Can Form Germ Cells *In Vitro. Proceedings of the National Academy of Sciences* **100(20):** 11457–11462.

Turnbull DM, Tuppen HA, Greggains GD, *et al.* (2010) Pronuclear Transfer in Human Embryos to Prevent Transmission of Mitochondrial DNA Disease. *Nature* **465:** 82–85.

V

van Balen F. (1998) Development of IVF Children. *Developmental Review* **18(1):** 30–46.

Vogel G, Holden C. (2007) Developmental Biology. Field Leaps Forward with New Stem Cell Advances. *Science* **318(5854):** 1224–1225.

W

Warren MA. (2005) *Moral Status: Obligations to Persons and Other Living Things.* Oxford University Press, Oxford.

Weissmann G. (2006) Science Fraud: From Patchwork Mouse to Patchwork Data. *The FASEB Journal* **20:** 587–590.

Wernig M, Meissner A, Foreman R, *et al.* (2007) *In Vitro* Reprogramming of Fibroblasts into a Pluripotent ES Cell-like State. *Nature* **448(7151):** 318–324.

Wertheimer A. (1999) *Exploitation.* Princeton University Press, Princeton.

Wilkinson S. (2007) Commodification. In: Ashcroft RE, Dawson A, Draper H, McMillan JR (eds), *Principles of Health Care Ethics.* John Wiley and Sons, Chichester, pp. 285–291.

Williams C, Kitzinger J, Henderson L. (2003) Envisaging the Embryo in Stem Cell Research: Rhetorical Strategies and Media Reporting of Ethical Debate. *Sociology of Health and Illness* **25:** 793–814.

Williams C, Wainwright SP, Ehrich K, Michael M. (2008) Human Embryos as Boundary Objects? Some Reflections on the Biomedical Worlds of Embryonic Stem Cells and Pre-Implantation Genetic Diagnosis. *New Genetics and Society* **27:** 7–18.

Wilmut I, Schnieke AE, McWhir J, *et al.* (1997) Viable Offspring Derived from Fetal and Adult Mammalian Cells. *Nature* **385(6619):** 810–813.

Woods S. (2008) Stem Cell Stories: From Bedside to Bench. *Journal of Medical Ethics* **34:** 845–848.

Y

Yu J, Vodyanik MA, Smuga-Otto K, *et al.* (2007) Induced Pluripotent Stem Cell Lines Derived from Human Somatic Cells. *Science* **318:** 1917–20.

Z

Zavos PM, Illmensee K. (2006) Possible Therapy of Male Infertility by Reproductive Cloning: One Cloned Human 4-cell Embryo. *Archives of Andrology* **52(4):** 243–254.

Zoloth L. (2002) Jordan's Banks: A View from the First Years of Human Embryonic Stem Cell Research. *American Journal of Bioethics* **2:** 3–11.

Lectures and Presentations

Barnett I, Steuernagel T. (7 April 2005) Framing Assisted Reproduction: Feminist Voices and Policy Outcomes in Germany and the U.S. Paper Presented at the Annual Meeting of The Midwest Political Science Association. Palmer House Hilton, Chicago, Illinois. Available at http://www.allacademic.com/meta/p86807_index.html

Steinberg A. (1 June 2010) Jewish Medical Ethics: May Humans Play God? (lecture), Oxford.

Newspaper Articles, Press Releases, Blogposts, and Online Resources

Akst J. (19 November 2009) 2nd Human hESC Trial? The Scientist. Available at http://classic.the-scientist.com/blog/display/56155/

Aldhous P. (2008) Are Male Eggs and Female Sperm on the Horizon? *New Scientist.* Available at http://www.newscientist.com/article/mg19726414.000-are-male-eggs-and-female-sperm-on-the-horizon.html

American Association for the Advancement of Science (18 December 2008) Cellular Reprogramming Leads Science List of Top 10 2008 Breakthroughs (press release). Available at http://www.aaas.org/news/releases/2008/1218breakthrough.shtml

Anon. (2 February 2008) Getting Ready for Same-sex Reproduction. *New Scientist.* Available at http://www.newscientist.com/article/mg19726413.000-editorial-getting-ready-for-samesex-reproduction.html

Anon (Stem Cell Institute). (20 May 2008) Veterinarians Achieve Success with Adult Stem Cell Therapy in Animals. Available at http://www.cellmedicine.com/vet-stem-cells/

Anon. (2008) A Stem Cell Success Story — UK Stem Cell Biologists Help to Deliver a New Bronchus for Claudia Castillo. *UK National Stem Cell Network Newsletter*, Winter: 8. Available at http://www.uknscn.org/downloads/newsletter_winter08.pdf

Anon. (21 October 2009) Transplanted Tissue Improves Vision: Study Shows Enhanced Visual Acuity. *Science Daily*. Available at http://www.sciencedaily.com/releases/2009/10/091021014628.htm

Anon. (3 February 2009) Chinese Researchers Make Cloned Human Blastocysts. *Cell News*. Available at http://cellnews-blog.blogspot.com/2009/02/chinese-researchers-make-cloned-human.html

Banjo S. (14 May 2010) Hedge-Fund Founder Bolsters Stem-Cell Research With $27 Million Gift. *Wall Street Journal*. Available at http://online.wsj.com/article/SB10001424052748704635204575242444135532502.html

Batty D. (17 May 2007) Hybrid Embryos Get Go-Ahead. *The Guardian*. Available at http://www.guardian.co.uk/science/2007/may/17/businessofresearch.medicineandhealth

California Institute for Regenerative Medicine (CIRM). (11 March 2010) CIRM Allocates $50 Million for Stem Cell Therapy Development, a Boost for the State's Growing Stem Cell Industry (press release). Available at http://www.cirm.ca.gov/PressRelease_031110

Carney C. (24 February 2007) Stem Cell Success for Spinal Injury in India. *Wired*. Available at http://www.wired.com/bodyhack/2007/02/stem_cell_succe/

Cohen CB, Brandhorst BP. (2 January 2008) Getting Clear on the Ethics of iPS Cells. *Bioethics Forum Blog*. Available at http://www.thehastingscenter.org/Bioethicsforum/Post.aspx?id=710

Connor S. (5 October 2009) Vital Embryo Research Driven Out of Britain. The Independent. Available at http://www.independent.co.uk/news/science/vital-embryo-research-driven-out-of-britain-1797821.html

Cressey D. (10 February 2010) Stem Cell Stroke Trial Gets Final Approval in UK. The Great Beyond. *Nature Blog*. Available at http://blogs.nature.com/news/2010/02/stem_cell_stroke_trial_gets_fi.html

Falco M. (21 Jan 2010) First U.S. Stem Cells Transplanted into Spinal Cord. *CNN*. Available at http://edition.cnn.com/2010/HEALTH/01/21/stem.cell.spine/index.html?utm_source=Full+List&utm_campaign=73859fdf92-SCOPE1_14_2010&utm_medium=email

Friend T. (11 July 2001) Group Creates Embryos Specifically for Research. *USA Today*. Available at http://cmbi.bjmu.edu.cn/news/0107/89.htm

Gray R. (27 Apr 2009) The Miracle Stem Cell Cures Made in Britain. *The Telegraph*. Available at http://www.telegraph.co.uk/technology/5232182/The-miracle-stem-cell-cures-made-in-Britain.html

Hall K. (3 February 2009) Japanese Stem Cell Scientist Announces Promising Results. *Business Week*. Available at http://www.businessweek.com/globalbiz/blog/eyeonasia/archives/2009/02/japanese_stem_cell_scientist_announces_promising_results.html?campaign_id=rss_blog_asiatech

Harris D. (12 October 2009) Groundbreaking Stem Cell Surgery Gives Boy New Cheekbones. *Good Morning America*. Available at http://abcnews.go.com/GMA/OnCall/experimental-treatment-boy-cheekbones/Story?id=8804636&page

Henderson M. (17 February 2010) Medical Potential of iPS Stem Cells Exaggerated Says World Authority. *The Times Online*. Available at http://www.timesonline.co.uk/tol/news/science/medicine/article7029447.ece

Henderson M. (2 April 2009) We Have Created Human–Animal Embryos Already, Say British Team. *The Times Online*. Available at http://www.timesonline.co.uk/tol/life_and_style/health/article3663033.ece

Henderson M. (23 December 2009) Stem Cell Eye Treatment Gives Victim of Fight His Sight Back. *The Times Online*. Available at http://www.timesonline.co.uk/tol/news/uk/health/article6965043.ece

Henderson M. (24 March 2008) Cloned Cells Bring Hope of Therapy for Parkinson's Disease. *The Times Online*. Available at www.timesonline.co.uk/tol/news/science/article3607659.ece

Human Fertilization & Embryology Authority (13 November 2008) *HFEA Chair Welcomes Royal Assent for HFE Act* (press release). Available at http://www.hfea.gov.uk/371.html

Human Genetics Alert 'GM and Human–Animal Hybrid Embryos — Confused About the Difference?' Available at http://www.hgalert.org/GM_and_hybrid_embryos_confused.html

Josephs L. (2 June 2010) Costa Rica shuts stem cell clinic. *Reuters*. Available at http://www.reuters.com/article/2010/06/02/us-costarica-stemcells-idUSTRE6516UR20100602

Karolinska Institutet. (2 February 2010) Stem Cells Rescue Nerve Cells by Direct Contact. *Science Daily*. Available at http://www.sciencedaily.com/releases/2010/02/100201171754.htm

Lister S. (30 January 2009) Stem Cell Therapy Reduces Symptoms of Multiple Sclerosis. *The Times Online*. Available at http://www.timesonline.co.uk/tol/life_and_style/health/article5614644.ece

Madeleine BLL. (22 June 2005) Embryonic Stem Cell Research: Accepting the Knowledge and Applying It to Our Lives. *Japan Inc. Communications.* Available at http://www.thefreelibrary.com/Embryonic+stem+cell+research:+accepting+the+knowledge+and+applying+it...-a0134293286

Martin J. (17 January 2008) Amen to Death of Embryo Research. *The Australian.* Available at http://www.theaustralian.com.au/news/opinion/amen-to-death-of-embryo-research/story-e6frg7ef-1111115333189

McArthur G. (30 May 2008) Miracle Man Says Embryonic Stem-Cell Therapy Works. *Herald Sun.* Available at http://www.heraldsun.com.au/news/victoria/miracle-man-takes-steps-to-recovery/story-e6frf7kx-1111116481636

Midgley S. (10 May 2008) Spine Injuries May Be Repaired by a Nose. *The Times Online.* Available at http://business.timesonline.co.uk/tol/business/specials/stemcell_research/article3904111.ece

Panizzo R. (12 April 2010) DNA Difference Between Stem Cell Types Found. *BioNews.* http://www.*BioNews*.org.uk/page_57804.asp

Pryor S. (22 February 2010) Fresh Fears over the Promise of iPS Cells. *BioNews.* Available at http://www.*BioNews*.org.uk/page_54791.asp

Roberts M. (18 March 2008) UK MPs Reconsider Artificial Gamete Ban. *BioNews.* Available at http://www.*BioNews*.org.uk/page_13333.asp

Sample I. (5 January 2007) Scientists attack plan to ban 'hybrid' embryos. *The Guardian.* Available at http://www.guardian.co.uk/science/2007/jan/05/medicalresearch.health

Science Timeline, Brits at Their Best. Available at http://www.britsattheirbest.com/ingenious/ii_21st_century.htm

Siemaszko C. (26 June 2009) $10,000 is an Egg-cellent Price, Says Stem Cell Panel. *Daily News.* Available at http://articles.nydailynews.com/2009-06-26/news/17925165_1_stem-cell-religious-groups

Templeton SK. (14 Sept 2009) Stem Cells Believed to Repair Damaged Hearts. *The Australian.* Available at http://www.theaustralian.com.au/news/world/stem-cells-believed-to-repair-damaged-hearts/story-e6frg6so-1225772469533

UK Stem Cell Bank. *Stem Cell Catalogue.* Available at http://www.ukstemcellbank.org.uk/stemcelllines/stemcellcatalogue.cfm

Winslow R, Mundy A. (23 January 2009) First Embryonic Stem-cell Trial Gets Approval from the FDA. *Wall Street Journal.* Available at http://online.wsj.com/article/SB123268485825709415.html

Woolf M. (December 30, 2007) IVF clinics destroy 1m 'waste' embryos. Available at http://www.timesonline.co.uk/tol/life_and_style/health/article3108160.ece

Reports & Guidelines

Academy of Medical Sciences. (2007) *Inter-species Embryos.* Available at http://www.acmedsci.ac.uk/p48prid51.html#downloads

Australian Government. (2005) Legislation Review: Prohibition of Human Cloning Act 2002 and the Research Involving Human Embryos Act 2002, Reports, Canberra, December 2005. Available at http://pandora.nla.gov.au/pan/63190/20060912-0000/www.lockhartreview.com.au/reports.html

Balen A. (February 2005) Ovarian Hyperstimulation Syndrome — A Short Report for the HFEA. Available at http://www.hfea.gov.uk/cps/rde/xbcr/hfea/OHSS_Report_from_Adam_Balen_2005(1).pdf

Canadian Institutes of Health Research. (2010) *Updated Guidelines for Human Pluripotent Stem Cell Research.* Available at http://www.cihr-irsc.gc.ca/e/42071.html

Commission of the European Communities. (2003) *Report on Human Embryonic Stem Cell Research* (European Commission, Brussels).

HM Government. (2007) Government Response to the Report from the Joint Committee on the Human Tissue and Embryos (Draft) Bill (Cm 7209). Available at http://www.dh.gov.uk/prod_consum_dh/groups/dh_digitalassets/@dh/@en/documents/digitalasset/dh_079145.pdf

HM Government. (2005) Government Response to the Report from the House of Commons Science and Technology Committee: Human Reproductive Technologies and the Law (The Stationary Office, Norwich). Available at http://www.dh.gov.uk/prod_consum_dh/groups/dh_digitalassets/@dh/@en/documents/digitalasset/dh_4117874.pdf

House of Commons Science and Technology Committee 'Inquiry into Human Reproductive Technologies and the Law' (Eighth Special Report of Session 2004–05). Available at http://www.publications.parliament.uk/pa/cm200405/cmselect/cmsctech/491/491.pdf

Human Fertilisation & Embryology Authority. (2006) *One Child at a Time.* Available at http://www.hfea.gov.uk/docs/One_at_a_time_report.pdf

Human Fertilisation & Embryology Authority. (2007) *Multiple Births and Single Embryo Transfer Review: Evidence Base and Policy Analysis* (HFEA (17/10/07) 401). Available at http://www.hfea.gov.uk/docs/AM_MB_and_SET_review_Oct07.pdf

Millbank J. (2002) *Meet the Parents: A Review on the Research of Lesbian and Gay Families.* Available at http://www.qahc.org.au/files/shared/docs/meet_the_parents.pdf

Mills P. (2004) *Human Fertilisation and Embryology Authority, Seed Review Consultation Proposals* (ELC (09/04) 02). Available at http://www.hfea.gov.uk/docs/ELC_SEED_Sept04.pdf

National Advisory Bioethics Commission. (1999) *Ethical Issues in Human Stem Cell Research*. NBAC, Rockville, MD.

National Institutes of Health. (2 December 2009) *First Human Embryonic Stem Cell Lines Approved for Use Under New NIH Guidelines*. Available at http://www.nih.gov/news/health/dec2009/od-02.htm

Pope John Paul II. (1995) *Evangelium Vitae. Papal Encyclical*. Available from http://www.vatican.va/edocs/ENG0141/_INDEX.HTM.

Joint Committee on the Human Tissue and Embryos (Draft) Bill. (2007) *Report of the Joint Committee on the Human Tissues and Embryos (Draft) Bill*. The Stationery Office, London.

Royal College of Physicians. (2007) Guidelines on the Practice of Ethics Committees in Medical Research with Human Participants, 4th Edition [10.13].

The Hinxton Group. (11 April 2008) *Consensus Statement: Science, Ethics and Policy Challenges of Pluripotent Stem Cell-derived Gametes*. Available at http://www.hinxtongroup.org/Consensus_HG08_FINAL.pdf

House of Lords Select Committee on Stem Cell Research. (2002) *Stem Cell Research*. Available at http://www.parliament.the-stationery-office.co.uk/pa/ld200102/ldselect/ldstem/83/8301.htm

Victoria Parliamentary Debates, Assembly (10 September 2008) 3442 (Robert Hulls, Attorney General).

Victorian Law Reform Commission. (2007) *Assisted Reproductive Technology & Adoption: Final Report*. Law Reform Commission, Melbourne.

Hansard

Powell E. (15 February 1985) House of Commons Debate, Vol. 73, cc637–702.

Legislation

England and Wales

Congenital Disabilities (Civil Liability) Act 1976
Human Fertilisation and Embryology Act 1990

Human Fertilisation and Embryology Act 1990 (as amended 2008)
Human Tissue Act 2004
Unborn Children (Protection) Bill 1985

Australia

Assisted Reproductive Technologies Act 2008 (Victoria)
Charter of Human Rights and Responsibilities Act 2006 (Victoria)
Prohibition of Human Cloning for Reproduction Act 2002
Research Involving Human Embryos Act 2002

Canada

Assisted Human Reproduction Act 2004

Case Law

United Kingdom

M R v T R & Ors [2006] IEHC 359 at 363
Parker v British Airways Board [1982] QB 1004, [1982] 1 All ER 834, CA
R v Herbert [1961] JPLGR 12, 13
R (Quintavalle) v Secretary of State for Health [2001] EWHC 918 (Admin)
Yearworth and Others v North Bristol NHS Trust [2009] EWCA Civ 37

European Union

Oliver Brüstle v Greenpeace (C-34/10) (2010/C 100/29)

Index